EMULSION POLYMERIZATION OF VINYL ACETATE

Papers based on a symposium on Emulsion Polymerization of Vinyl Acetate held between 10 and 11 April 1980 at the Emulsion Polymers Institute, Lehigh University, Bethlehem, Pennsylvania, USA

Emulsion Polymerization of Vinyl Acetate

Edited by

MOHAMED S. EL-AASSER

and

JOHN W. VANDERHOFF

Co-directors,
Emulsion Polymers Institute, Lehigh University,
Bethlehem, Pennsylvania, USA

APPLIED SCIENCE PUBLISHERS
LONDON and NEW JERSEY

APPLIED SCIENCE PUBLISHERS LTD
Ripple Road, Barking, Essex, England

APPLIED SCIENCE PUBLISHERS, INC.
Englewood, New Jersey 07631, USA

British Library Cataloguing in Publication Data

Emulsion polymerization of vinyl acetate.
1. Vinyl polymers – Congresses
2. Polymers and polymerisation – Congresses
I. El-Aasser, Mohamed S.
II. Vanderhoff, John W.
547'.28 QD281.P6

ISBN 0-85334-971-1

WITH 65 TABLES AND 108 ILLUSTRATIONS

The selection and presentation of material and the opinions expressed in this publication are the sole responsibility of the authors concerned.

Typeset by Unicus Graphics Ltd, Horsham
Printed in Great Britain by Galliard (Printers) Ltd, Great Yarmouth

Introduction

It is particularly appropriate that this symposium on the emulsion polymerization of vinyl acetate was held in recognition of the industrial importance of poly(vinyl acetate) and vinyl acetate copolymers, and their rather unique properties among emulsion polymers in general. Poly(vinyl acetate) latexes were the first synthetic polymer latexes to be made on a commercial scale: their production using polyvinyl alcohol as emulsifier began in Germany during the mid-1930s and has continued to the present day, growing steadily with the years. Indeed, poly(vinyl acetate) latexes prepared with polyvinyl alcohol are still one of the mainstays of the adhesives industry. With the passing of time, however, vinyl acetate copolymers have been developed: copolymers with maleate esters such as dibutyl maleate, acrylate esters such as ethyl acrylate and butyl acrylate, versatic acid esters, and, more recently, ethylene. These versatile copolymers have found increasing use in more sophisticated adhesives with specialized properties, adhesives for clay coatings on paper, carpet backing, and interior and exterior paints. Thus more than 45 years after the first commercial production of vinyl acetate latexes, their use is still growing, both in actual quantities and different applications.

The industrial importance of vinyl acetate latexes makes the mechanism and kinetics of their emulsion polymerization of practical as well as scientific interest. However, little work was done on this subject outside industrial circles, and although vinyl acetate latexes were the first to be produced commercially, it was the butadiene–styrene copolymer latexes that provided the stimulus for development of the mechanism and kinetics of emulsion polymerization. The forced development of these copolymers in the Office of Rubber Reserve Program of World War II, not only resulted in the rapid development of emulsion polymerization processes to make synthetic rubber, but also gave the first coherent account of the mechanism by emulsion polymerization that produces latex polymer particles of *ca.* 0·1 μm from monomer droplets of 1–10 μm. This was the concept of initiation of polymerization in monomer-swollen emulsifier micelles advanced by Harkins[1] and developed quantitatively by Smith and Ewart.[2] This latter theoretical development, with its successful use of the Smith–Ewart Case 2 in explaining the emulsion polymerization of styrene, stimulated research in many other monomers. It was soon found that the Smith–Ewart Case 2 fitted the emulsion polymerization of styrene in only a narrow range of conditions and fitted the emulsion polymerization of other monomers even more poorly. However, the stimulus provided by the simplicity of the resulting kinetic equations resulted in the investigation of many other monomers, including vinyl acetate.

The early workers (French[3] and O'Donnell *et al.*[4]) interpreted their results for vinyl acetate in terms of the Smith–Ewart Case 2, but their conversion–time curves were determined gravimetrically, with relatively few points which could be well represented by a straight line. Thus these data were interpreted in accord with the initiation-in-micelles hypothesis despite some evidence to the contrary in industrial circles. Certainly, the traditional polyvinyl alcohol emulsifier did not lend itself well to interpretation in terms of concentrations above and below the critical micelle concentration.

Soon, however, Priest[5] pointed out for vinyl acetate, as did Jacobi[6] for vinyl chloride, that initiation in micelles did not fit the experimental evidence and proposed that initiation occurred in the aqueous phase. Initiation in the aqueous phase was not considered important in the emulsion polymerization of styrene, but vinyl acetate is about 100 times more soluble in water than styrene, so that a primary radical formed in the water phase could add monomer units much more readily. These oligomeric radicals thus formed would grow until they reached their solubility limit in water and then precipitate to nucleate a latex particle. In some cases, these particles upon precipitation would be destabilized and flocculate with other such particles or with particles formed earlier.

This concept was expanded and confirmed experimentally by Stannett *et al.*,[7] Napper and Alexander,[8] and Dunn and Tylor.[9]

At the same time, the hypothesis of initiation in the adsorbed emulsifier layer was developed by Medvedev,[10] and some time later the hypothesis of initiation in small monomer droplets was proposed.[11] At one time or another, all four of these mechanisms of particle initiation—monomer-swollen micelles, aqueous phase, adsorbed emulsifier layer, and monomer droplets—have been proposed to explain the experimental results obtained for vinyl acetate.

Indeed, vinyl acetate differs from other monomers in other respects than its solubility in water: it has a high equilibrium monomer–polymer swelling ratio (values as great as 7 : 1), it has a high transfer constant (*ca.* 10^{-3} as compared to *ca.* 10^{-4} for styrene), and its radical has a strong tendency to abstract hydrogens from previously formed poly(vinyl acetate) to form branches.

The poly(vinyl acetate) also differs from other polymers. Although generally hydrophobic, it shows some hydrophilic characteristics. It can be solubilized in emulsifier solutions, a process that is still not well understood and whose significance at the earlier stages of emulsion polymerization has not yet been fully taken into account. Moreover, in the presence of water, it can hydrolyze to form vinyl alcohol units, which makes the poly(vinyl acetate) particle, slightly hydrolyzed in this way, somewhat self-stabilizing in the absence of emulsifier.

Papers presented on these various aspects of vinyl acetate emulsion polymerization and the properties and behavior of poly(vinyl acetate) latexes have usually been presented in symposia and publications dealing with other types of emulsion polymers. Because these properties are significantly different from those of other emulsion polymers and because poly(vinyl acetate) has such strong industrial importance, this symposium was organized to gather together those who are working in this field, to exchange ideas, to review progress, to formulate hypotheses, and to determine areas of agreement as well as areas of disagreement; it is fitting that a symposium was held solely to discuss those unique properties and problems of vinyl acetate. It is hoped that the papers presented at the symposium, which are collected in this volume, will indeed represent the current thinking on vinyl acetate systems and furnish a starting point for the development of a unified theory which will reconcile all parameters of vinyl acetate emulsion polymerization with those of other monomers.

J. W. Vanderhoff
M. S. El-Aasser

REFERENCES

1. W. D. Harkins, *J. Am. Chem. Soc.*, **69**, 1428 (1947).
2. W. V. Smith and R. H. Ewart, *J. Chem. Phys.*, **16**, 592 (1948).
3. D. M. French, *J. Polym. Sci.*, **32**, 395 (1958).
4. J. T. O'Donnell, R. B. Mersobian and A. E. Woodward, *J. Polym. Sci.*, **28**, 171 (1958).
5. W. J. Priest, *J. Phys. Chem.*, **56**, 1077 (1952).
6. B. Jacobi, *Angew. Chem.*, **64**, 539 (1952).

7. V. T. Stannett, M. Litt and R. Patsiga, *J. Phys. Chem.*, **64**, 801 (1960).
8. D. H. Napper and A. E. Alexander, *J. Polym. Sci.*, **61**, 127 (1962).
9. A. S. Dunn and P. A. Tylor, *Makromol Chem.*, **83**, 207 (1965).
10. S. S. Medvedev, *Proc. Int. Symp. Macromolek. Chem., Prague, 1957*, London and New York, Pergamon Press, p. 174.
11. J. Ugelstad, M. S. El-Aasser and J. W. Vanderhoff, *J. Polym. Sci., Polym. Letts.*, **11**, 503 (1973).

Contents

List of Contributors

S. M. Ahmed
Hercules Inc., Research Center, Wilmington, Delaware, USA

T. Bone
Celanese Plastics and Specialties Company, Jeffersontown, Kentucky 40299, USA

Nicoleta Carp
Centre for Physical Chemistry, Splaiul Independenţei 202, Bucharest 77208, România

R. R. Challa
Department of Chemical Engineering, North Carolina State University, Raleigh, North Carolina 27650, USA

K. H. S. Chang
Department of Macromolecular Science, Case Western Reserve University, Cleveland, Ohio 44106, USA

W. E. Daniels
Air Products and Chemicals Company, Box 538, Allentown, Pennsylvania 18105, USA

I. Diaconescu
Centre for Plastic Materials Research, Splaiul Independenţei 202, Bucharest 77208, România

D. Donescu
Centre for Plastic Materials Research, Splaiul Independenţei 202, Bucharest 77208, România

J. H. Drew
Department of Chemical Engineering, North Carolina State University, Raleigh, North Carolina 27650, USA

A. S. Dunn
Chemistry Department, University of Manchester Institute of Science and Technology, Sackville Street, Manchester M60 1QD, UK

M. S. El-Aasser
Emulsion Polymers Institute, Department of Chemical Engineering, Lehigh University, Bethlehem, Pennsylvania 18015, USA

C. T. Enos
Air Products and Chemicals Company, Box 538, Allentown, Pennsylvania 18105, USA

J. Frost
Air Products and Chemicals Company, Box 538, Allentown, Pennsylvania 18105, USA

C. Gajria
Celanese Plastics and Specialties Company, Jeffersontown, Kentucky 40299, USA

Kristiana Goşa
Centre for Plastic Materials Research, Splaiul Independenţei 202, Bucharest 77208, România

A. E. Hamielec
Department of Chemical Engineering, McMaster University, Hamilton, Ontario, Canada

J. G. Iacoviello
Air Products and Chemicals Company, Box 538, Allentown, Pennsylvania 18105, USA

A. Klein
Emulsion Polymers Institute, Department of Chemical Engineering, Lehigh University, Bethlehem, Pennsylvania 18015, USA

M. H. Litt
Department of Macromolecular Science, Case Western Reserve University, Cleveland, Ohio 44106, USA

M. F. Llauro
Laboratoire des Materiaux Organiques, CNRS, Villeurbanne, France

T. Makgawinata
Emulsion Polymers Institute, Department of Chemical Engineering, Lehigh University, Bethlehem, Pennsylvania 18015, USA

Magdalena Mazăre
Centre for Plastic Materials Research, Splaiul Independenţei 202, Bucharest 77208, România

S. Misra
Emulsion Polymers Institute, Department of Chemical Engineering, Lehigh University, Bethlehem, Pennsylvania 18015, USA

M. Nomura
Department of Macromolecular Science, Case Western Reserve University, Cleveland, Ohio 44106, USA

C. Pichot
Laboratoire des Materiaux Organiques, CNRS, Villeurbanne, France

E. P. Stahel
Department of Chemical Engineering, North Carolina State University, Raleigh, North Carolina 27650, USA

V. T. Stannett
Department of Chemical Engineering, North Carolina State University, Raleigh, North Carolina 27650, USA

J. W. Vanderhoff
Emulsion Polymers Institute, Department of Chemistry, Lehigh University, Bethlehem, Pennsylvania 18015, USA

B. R. Vijayendran

Celanese Plastics and Specialties Company, Jeffersontown, Kentucky 40299, USA

Henry Warson

Solihull Chemical Services, 284 Warwick Road, Solihull, West Midlands B92 7AF, UK

Richard L. Zollars

Department of Chemical Engineering, Washington State University, Pullman, Washington 99164, USA

1

Frontiers Between Theory and Industrial Practice in Vinyl Acetate Polymerizations

HENRY WARSON
Solihull Chemical Services, Solihull, West Midlands, UK

SUMMARY

Vinyl acetate is not polymerized commercially in dilatometers, but the application of theory to practical emulsion polymerization helps our understanding of the processes–this is the theme of this chapter.

SPECIFIC PROBLEMS OF INDUSTRIAL PRODUCTION

These may be listed as follows, and it should be noted that many, except possibly precipitation, are not self-evident in theoretical work.

(1) Exact repeatability within narrow limits is desirable; this may refer to inhibition time, if any; the final solids content should be constant ± 1%, and any residual monomer maintained within narrow low limits; likewise particle size, emulsion viscosity and, probably, polymer viscosity (i.e. molecular weight average) should vary little from batch to batch. This leads to:
(2) The formulation should be robust, i.e. not subject to variation with minor changes of raw material or of operative conditions, and:
(3) The process should be capable of operation by a skilled, but non-graduate plant worker who is trained to the job.
(4) Further to (2), a formula should give a reasonably satisfactory latex even with substitute raw materials which are not entirely satisfactory, e.g. the polyvinyl alcohol available might not be the best grade, or an alkylbenzene sulfonate may be heavily loaded with filler. This may occur in 'developing countries'.

(5) Raw materials should be useable as supplied; if monomer requires distilling, a simple operation should be adequate, or a single wash with an alkaline solution to remove an inhibitor such as hydroquinone. Preferably monomers should be useable as supplied.

The following general conditions with regard to the process should be noted:

(6) The process should be complete within a reasonable time, an overall time including loading, cooling and unloading should be 8–16 hours, but may be about 24 hours where dienes are included and it is desirable to finish the reaction rather than to distill excess unpolymerized monomer. In some cases steam, air or nitrogen may be needed to remove final traces of monomer for safety or odour reasons, e.g. vinyl chloride where this is a comonomer. Normally with a vinyl acetate (co)polymerization, the reactions are carried to substantial completion.
(7) It is normally desirable to prepare a poly(vinyl acetate) latex at the highest concentration possible, usually >50%, unlike theoretical work. The nature and shape of the stirrer may affect performance—usually this is determined empirically. Sometimes baffles are desirable to improve mixing, especially with viscous and thixotropic latexes; the vessel shape may be significant—cylindrical, with the height about double the diameter is an average shape—squat wide reactors may give heat problems unless solved by intense reflux.
(8) Excessive foaming should be avoided, but some anti-foam agents may cause cissing when a latex is dried to a film.
(9) It is necessary to avoid excessive or irregular precipitation either in the reactor or on the reactor walls and stirrer. Polyvinyl alcohol stabilized latexes are usually the most difficult in this respect; excess 'baking' of the sides of the reactor should be avoided. If present, however, precipitate should be easily filtered, or removable from the walls and stirrer, possibly by alkaline action.

LACK OF APPRECIATION BY INDUSTRIAL LABORATORIES OF THEORY

This may include the following.

(1) A lack of appreciation of statistical experimental batch to batch variations, and often of scaling up problems.
(2) A large number of unnecessary empirical experiments, e.g. in the copolymerization of vinyl acetate with acrylic esters (see page 6).

(3) Failure to perform any fundamental work, e.g. on copolymerization, the effects of different emulsifiers or initiators, and failure to give any consideration to such fundamental factors as the nature of particle formation, the theory of which has been developed considerably.

Let me give some historical examples based on my own experience. One factor which disturbed early work in batch processes was the failure of any batches made with hydrogen peroxide to give satisfactory and repeatable results–these were only obtained with a persulfate initiator. We now know that sulfate or sulfonate end groups are formed on poly(vinyl acetate) molecule ends which are self-stabilizing. I believe that in redox processes they are most likely to be sulfonate types, from the reductant, although the contrary has been claimed.

One of the earliest successful water soluble polymers used as a stabilizer for poly(vinyl acetate) emulsions was gum arabic (acacia), and I believe that it is still used by some manufacturers. It has been my personal 'bête noire'. In many cases successful laboratory processes have failed unpredictably and irregularly on a production scale, with heavy precipitation. It is known that after a complete polymerization with this polymeric stabilizer there is residual poly(vinyl acetate) which is insoluble in any standard solvent, although the entire latex can apparently be dissolved by careful treatment with ethanol/benzene in selected ratios. Although this work is at least 35 years old, I know of no explanation of the unique function of this natural polymer. It is known that there are large annual variation in crops–possibly there may be some grafting.

POLYVINYL ALCOHOL

Over 30 years ago difficulty was caused by variability of the 'polyvinyl alcohol'. A fairly low molecular weight polymer, consisting of 88% hydrolyzed poly(vinyl acetate) was the most desirable, but laboratory and some larger scale emulsions showed highly variable results in polymer particle size, and emulsion viscosity and stability. Thus when one polyvinyl alcohol which was suitable for a process was replaced by another, reputedly of better quality, from the same manufacturer, the process failed completely from the stability aspect. Many years later it was disclosed that the original product was heterogeneous in that it had a wide scatter of both molecular weights and molecular species, whilst the 'improved' product was a sharp 'cut' on both aspects.

Polyvinyl alcohol stabilized latexes are amongst the principal classes, and understanding of the stabilizing action depends on understanding the nature of polyvinyl alcohol and its behaviour during polymerization. Thus 'blocky' polymers in which the residual acetate groups tend either to concentrate at one end of the polymer chains or in blocks along their length are considered to produce more stable latexes than those with a random distribution of acetyl groups. The process of preparing a polyvinyl alcohol by hydrolysis of an emulsion with a sulfonic acid[1,2] or of a copolymer of vinyl acetate and a hydrolysis resistant monomer[3,4] produces polyvinyl alcohols in which the hydrophobic groups should have random distribution and produce stable poly(vinyl acetate) latexes when used as emulsifiers, but these may give extra stabilization due to acid end groups (as below).

The following variations in polyvinyl alcohol have been summarized, and it is stressed that this refers substantially to a poly(vinyl acetate), partially hydrolyzed, usually about 88%.

(1) Variations in scatter of molecular weight.
(2) Variation in scatter of molecular species, i.e. the product may vary from almost fully hydrolyzed polymer to only slightly hydrolyzed polymer, giving an apparent average.
(3) Slight variation of head-to-head polymer content as distinct from the normal head-to-tail copolymer.
(4) Variations in chain branching due to the manufacture of the original poly(vinyl acetate).
(5) Variations in tacticity of the polymers, although this is only likely to be noted in special cases, e.g. a polyvinyl alcohol from the trifluoroacetate.
(6) Variations in distribution of the acetate and hydroxyl groups along the chain dependent on the manner of preparation. In alkaline processes there is probably a tendency to chain 'stripping' and blockiness. Acid processes should give a more even distribution of hydroxyl and acetate groups, and in some cases may have end sulfate/sulfonate groups.
(7) Variations due to end groups from the original polymerization initiator.
(8) Presence of fortuitous groupings, e.g. ketonic or carboxylic groups due to oxidation.
(9) Unchanged monomer in the initial poly(vinyl acetate) may lead to the formation of acetaldehyde by hydrolysis, and hence the forming of some acetal groups on the polyvinyl alcohol.[5]

In addition to the above there is a further possibility:

(10) The hydroxyl groups may possibly act as the redox components in a redox initiation.

GRAFTING

I suggest that investigations into grafting of vinyl acetate on to polyvinyl alcohol, of which only a little fundamental research has been attempted, would be rewarding. In a practical delayed-addition-of-monomer process, where there is little free monomer at any one time because of a slow rate of addition, there is a much greater tendency for monomer to become initiated or to react by addition in the water phase or on the surface of the particles of polymer, and there is obviously a much greater tendency to graft on to the stabilizer. There is little doubt that batch to batch variations in emulsion viscosity and sometimes stability are due to grafting on to a polyvinyl alcohol emulsifier. Excessive grafting would obviously cause instability since the overall molecules of polyvinyl alcohol are not sufficiently hydrophilic after grafting to provide stability unless some ionic groups have been introduced.

The process of extensive graft in emulsion polymerization probably occurs when the 'Pluronic' type of surfactant is used, it being difficult to account for the properties of the resultant emulsion otherwise.

The gradual addition of monomers, polyvinyl alcohol solution and a redox initiator to a 'seed' of preformed emulsion produces a highly branched polymer, probably by extensive grafting in which some intramolecular cross-linking occurs resulting in the latex being tolerant to addition of ethyl alcohol, and with the property of being very useful in adhesive formulations.

FREE MONOMER ANALYSIS

The analysis of free monomer at any one time is desirable if a process is to be properly controlled. Apart from the grafting possibilities already mentioned, the rate of addition of monomer in the standard gradual addition types may affect the number of nuclei formed, and hence both the ultimate particle size, and scatter of particle size. This is not so serious with a vinyl acetate polymerization as with the more insoluble monomers such as styrene or 2-ethylhexyl acrylate.

A rapid method of testing is essential for residual monomer so that, if necessary, the rates of addition of monomers, or sometimes of additional water phase, can be adjusted. Normal bromination methods are much too slow, and whilst one alternative method, that of Das (mercuric acetate reaction) is possible, the only truly rapid method of ascertaining residual monomer during an emulsion polymerization is by gas liquid chromatography, which also has the advantage that it will show all monomers, or extraneous moderately volatile matter, as separate peaks on the chart. It is also highly useful in estimating unchanged monomer at the conclusion of a polymerization.

The importance of estimating monomers during a copolymerization has been generally overlooked. Thus when vinyl acetate is copolymerized with an acrylic ester because of the differing reactivities, unless either the addition of mixed monomers is restricted so that there is very little free monomer at any one time or the monomers are added in different streams, a large amount of almost pure poly(vinyl acetate) will form at the end of the reaction, whilst that formed earliest is richest in acrylic ester units. As the compatibilities of the corresponding homopolymers decrease with increasing length of the alkyl ester chain of the acrylate ester, the properties of these mixtures will differ from those of true copolymers. Many processes ignore this simple fact resulting in incorrect statements in earlier literature, e.g. that butyl acrylate is a better 'internal' plasticizer than 2-ethylhexyl acrylate.

On the other hand it may be desirable to form a heterogeneous copolymer, e.g. for vibration damping purposes, and in this case residual monomer checks are desirable during a polymerization.

It is interesting to note that if the vinyl ester of the highly branched synthetic acid known as 'Versatic acid' is copolymerized with vinyl acetate it is desirable to hold about 5% of the vinyl acetate until a late stage in the reaction. Because of the use of water soluble initiators and the water solubility of vinyl acetate, it tends to polymerize somewhat faster than the comonomer ester.

KINETICS

With regard to the kinetics in emulsion it must have been obvious from the start that because of the appreciable aqueous solubility of vinyl acetate, and also because both monomer and many surfactants tend to solubilize poly(vinyl acetate) to some extent, that Smith–Ewart kinetics would not be valid for vinyl acetate polymerization. Some of the other chapters discuss the subject; I favour ideas of aqueous initiation with propagation in both water and monomer phases.

MECHANISMS OF STABILIZATION

The fundamental work on the causes of stabilization, and the best method of achieving this owes much to Lehigh University and John Vanderhoff. Theories of adsorption on the surface of particles do much to explain stability, and in some cases the lack of it. The correct hydrophobe–hydrophile balance, which may be different in a copolymer, is necessary. Thus fully hydrolyzed polyvinyl alcohol is unsatisfactory as a stabilizer in most circumstances.

In addition, the theory which I find most satisfactory is developed from the ideas of Schulman and Cockbain.[6] This is the idea of a condensed film on the surface of the particles. Thus a pure petroleum fraction (Nujol) did not remain in a stable emulsion with sodium lauryl sulfate, but was quite stable in the presence of equimolar proportions of sodium lauryl sulfate and cetyl alcohol or cholesterol. This led, as early as 1945, to the observation that by this principle stable emulsions of polymethyl methacrylate could be prepared which would be unstable without the hydroxyl compound.[7]

The development has been greatly extended by J. Ugelstad and co-workers with special reference to vinyl chloride and styrene, but under some conditions operates for vinyl acetate latexes; $1\frac{1}{2}\%$ of cetyl alcohol with 4% of a polyvinyl alcohol (12% residual acetate molar) and a fractional percentage of a surfactant gives a high stability latex.

The principle can be extended however; and if in general there is any marked difference in the hydrophobe–hydrophile characteristics of two emulsifiers, examples being the nonionic polyoxyethylene ethers with different oxyethylation degrees, stability of a latex is usually greater than if only one emulsifier with the average number of oxyethylene groups is included. In this case a variation in commercial materials is easily understood. Thus a polyoxyethylene ether with a broad spectrum of ethylene oxide groups may well function better than one with a sharp cut. It is simple to envisage a condensed film on the surface formed by two polyoxyethylene ethers of quite different hydrophobe–hydrophile balance.

PRECIPITATON DURING POLYMERIZATION

Poly(vinyl acetate) latexes are now almost invariably produced by a delayed addition process, in which 0–10% is added at the start, and the balance gradually. In some cases there is also a gradual addition of water phase or water phase components. In most cases, with the exception of those which are stabilized with polyvinyl alcohol, if an attempt is made to prepare a stable latex in one stage at much above 40%, some precipitation of polymer occurs during the reaction. Why is this?

There are probably several simultaneous processes. Poly(vinyl acetate) is known to absorb all free monomer at a quite low level of polymerization, about 15%.[8] The particles are then in what is unscientifically referred to as 'the sticky state'—a concentrated solution of polymer in monomer. It is natural therefore that coalescence will occur unless there is a strong surface barrier on the particles as occurs with a polyvinyl alcohol stabilizer. Equally, there may be too many particles formed at an early stage for the emulsifier present to give stability. During a one-stage emulsion polymerization with a persulfate initiator the particle size tends to reduce. Ignoring size reduction by shear, this shows that polymer particles form independently of the monomer particles. As they grow there is a tendency for the surface to be reduced to maintain stability. Once an unstable particle, i.e. a precipitate, is formed, it acts as an automatic nucleus for further precipitation.

There is thus a theoretical basis for attaining stability if a one-stage emulsion polymerization was accompanied by a gradual addition of emulsifier, just sufficient to maintain stability as the particle size increases without creating new nucleation. Particle size increase may also occur by limited fusion of smaller particles.

The delayed addition procedure avoids the problem, and provided that there is no accumulation of monomer and that the stabilizer present is adequate to prevent particle coalescence—this depends on its physical state as much as complete coverage—the intermediate 'sticky state' is avoided, and thus greater stability is attained.

It is highly probable that self-emulsifying monomers of the type of sodium sulfoethyl methacrylate $NaO_3SCH_2OOCC(CH_3){:}CH_2$ by providing charged particles, would materially assist stability.

CATIONIC LATEXES

These have been generally difficult to prepare with vinyl acetate homopolymers, although they seem more satisfactory with copolymers. The problem is probably oxidation products of cationic emulsifiers which may react with vinyl acetate or acetaldehyde. The problem may be overcome by the use of the cationic initiator 2,2′-azobis(2-amidopropane) hydrochloride.[9]

COPOLYMERS

Some of the problems have already been considered, e.g. the copolymerization of vinyl acetate with acrylic esters. Much early work, particularly in connection with technical service applications such as emulsion paints, completely ignored the formation of very heterogeneous polymer mixtures. Note that with 2-ethyexyl acrylate and vinyl acetate $r_1 : r_2$ is approximately 60 : 1 in favour of the acrylic ester.

When vinyl acetate is polymerized in a weight ratio of about 3 : 1 in emulsion with VV10 (Veova 10 in Europe), this being the vinyl ester of a C_{10} branched chain carboxylic acid in which the C to carboxyl is quaternary, it is found in practice that there is always a slight excess of the C_{10} monomer at the end, although $r_1 : r_2$ is approximately 1:1. The VV10, on film formation, leaves a prolonged after-odour, making the otherwise good latex useless as the basis of an emulsion paint. I believe this is due to the fact that as we normally use water soluble initiators, an appreciable part of the initiation and propagation is in the water phase, and probably the absorption of VV10 into the polymer phase is much slower than with vinyl acetate. This disturbs the normal equilibrium in favour of vinyl acetate. Practically, the problem is overcome simply by holding back 3–5% of vinyl acetate until polymerization is substantially complete. This ensures that the traces of VV10 are completely polymerized.

ETHYLENE COPOLYMERS

By considering ethylene as a liquid at the corresponding density and allowing for aqueous solubility, which can be increased with the addition of *tert*-butanol, it should be possible to copolymerize vinyl acetate with ethylene at about 11 kg/mm^2, the order of pressure at which butadiene is polymerized. The reactivity ratios are comparable, and by the gradual addition of vinyl acetate under pressure, the normal type of ratios of the two monomers should be obtained in a copolymer without difficulty.[10]

REFERENCES

1. J. E. O. Mayne and H. Warson (Vinyl Products), British Patent 655734, 1951.
2. J. E. O. Mayne, H. Warson and R. M. Levine (Vinyl Products), British Patent 766565, 1957.
3. H. Warson and H. Robertson (Dunlop), British Patent 1174097, 1969.
4. H. Warson, in *Properties and Applications of Polyvinyl Alcohol,* A. Finch (ed.), Society of Chemical Industry, London, 1968, pp. 46–76.
5. H. Warson, in *Ethylene and its Industrial Derivatives,* S. A. Miller (ed.), Ernest Benn, London, 1969, pp. 980–2.

6. J. H. Schulman and E. G. Cockbain, *Trans. Faraday Soc.,* **36,** 651 (1940).
7. J. E. O. Mayne, H. Reichard and H. Warson (Vinyl Products), British Patent 607704, appl. 1945.
8. D. M. French, *J. Polym. Sci.,* **32,** 395 (1958).
9. K. Yamazaki *et al.* (Kao Soap), Japanese Patent 77-53,086.
10. Borden, German Patent 2,045,331, 1971.

2

The Polymerization of Aqueous Solutions of Vinyl Acetate

A. S. DUNN
Chemistry Department, University of Manchester Institute of Science and Technology, UK

SUMMARY

Since the difference between the characteristics of the emulsion polymerization of vinyl acetate and those of styrene largely arises from the greater water solubility of vinyl acetate, the polymerization of aqueous solutions of vinyl acetate was investigated. Because poly(vinyl acetate) is precipitated and the monomer is mostly extracted into the polymer phase, the reaction proved to be essentially an emulsion polymerization in Interval III. When persulphate is used as initiator, stable, monodisperse latexes can be produced without the use of an emulsifier. The use of anionic or polymeric (i.e. polyvinyl alcohol) emulsifiers stabilizes smaller latex particles. Much larger particles are produced when hydrogen peroxide initiation is used. Minor structural characteristics of polyvinyl alcohol affect the course of the reaction profoundly. A number of previously unpublished results relating to the effect of ionic strength on the polymerization, solubilization of vinyl acetate by emulsifiers, and the effect of vinyl acetate on the density of water are incorporated in this review.

The importance of the emulsion polymerization of vinyl acetate derives from the fact that poly(vinyl acetate) copolymers which contain a major proportion of vinyl acetate have found increasing acceptance as the film forming resin in emulsion paints and latex adhesives over the last 40 years. These materials were first developed in Germany during the 1939–45 war and were originally disclosed in the BIOS (British Intelligence Objectives Subcommission) and the US FIAT (Field Intelligence Agency Technical) reports made subsequently by teams of scientific investigators who had the task of finding out what developments had been made in German industry during the war. Manufacture of emulsion paints based on poly(vinyl acetate) latex was begun in England in 1948 on the basis of this information and most emulsion paint sold in Europe is of this type. Events took a different course in North America because of the need to find alternative applications for the output of the synthetic rubber plants once supplies of natural rubber became available again, and, consequently, latex paints based on butadiene–styrene copolymers dominated the market for many years: synthetic rubber was not produced in Britain until International Synthetic Rubber Ltd was set up by a consortium of oil companies in 1958. By the time Bovey *et al.*'s *Emulsion Polymerization*[1] was published in 1955, most (but not all) of the results of the crash programme of research undertaken under the auspices of the Office of the Rubber Reserve in the United States during the war into emulsion polymerization for the production of butadiene–styrene and butadiene–acrylonitrile synthetic rubbers had been published and these results are well summarized in that book. Because butadiene is a gas, it is much more convenient simply to investigate the emulsion polymerization of styrene which is widely used as a model monomer, although the homopolymerization of styrene in emulsion is of very limited industrial significance. The information available on the emulsion polymerization of styrene was very adequately systematized in the qualitative theory published by Harkins[2] in 1945 and later, and by the quantitative theory given by Smith and Ewart[3] in 1948, particularly when it was shown by Bartholome *et al.* in 1956[4] and by Burnett and Lehrle in 1959[5] that the propagation rate constant, k_p, for styrene derived from measurements of rates of emulsion polymerization by the use of Smith and Ewart's theory was in good agreement with values derived from rates of bulk polymerization.

EFFECT OF WATER SOLUBILITY OF MONOMERS

Thus it might have appeared even on a careful assessment of the situation in 1959 that the topic of emulsion polymerization was largely worked out (although subsequent developments have shown that this was not so) but for the suspicion that the theory which had been found so satisfactory for styrene did not really fit the facts for the other monomers, vinyl acetate and vinyl chloride, of which the emulsion polymerization was of major commercial importance. The discrepancies appeared to be attributable to the greater water solubility of vinyl acetate (2·8% at 60°C compared with 0·054% for styrene) and the insolubility of polyvinyl chloride in its monomer. Conse-

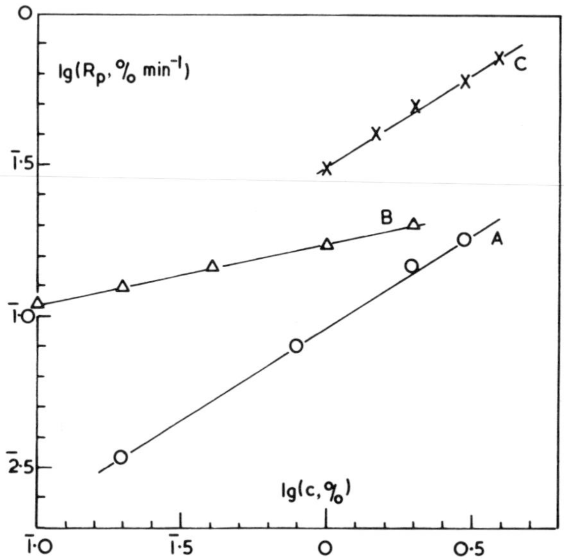

Fig. 1. Logarithmic plot showing dependence of order of reaction in emulsifier on the solubility of the monomer in the continuous phase. (A) Styrene in water: order 0·6. (B) Styrene in aqueous methanol: order 0·2. (C) Vinyl hexanoate in water: order 0·6.[8] (Redrawn from the *Journal of Polymer Science,* by permission).

quently it was decided to investigate the polymerization of aqueous solutions of vinyl acetate in the hope that this would show why the behaviour of this monomer differed from that of styrene in emulsion polymerization. The results of investigations on the polymerizations of aqueous solutions of methyl methacrylate[6] and acrylonitrile[7] were already available. The suggestion that the greater water solubility of vinyl acetate was important soon received convincing confirmation from the work of Okamura and Motoyama.[8] Figure 1 shows that vinyl hexanoate (i.e. vinyl caproate) which has a solubility in water similar to that of styrene but which has a transfer constant to monomer similar to that of vinyl acetate does have an order of reaction with respect to emulsifier concentration of 0·6, the same as for styrene which has a much lower transfer constant to monomer. On the other hand if the solubility of styrene in the continuous phase is increased by addition of alcohol, the order of reaction in emulsifier is reduced to a value similar to that for vinyl acetate. Nevertheless, subsequent work has shown that the value of the transfer constant is also a critical factor in determining whether or not monomers conform to Case 2 of the Smith–Ewart theory (in which the average number of radicals per latex particle, $\bar{n} = \frac{1}{2}$). Methyl methacrylate for which the solubility in water is about half of that of vinyl acetate, which is still very much higher than that of styrene or vinyl octadecanoate, but which has a low transfer constant to monomer similar to that of styrene, does conform to Case 2.[9] Vinyl acetate and vinyl chloride which have high transfer constants do not; and in fact it has recently been recognized[10,11] that it is Smith and Ewart's Case 1 (in which the transfer of radicals out of latex particles is not negligible so that consequently $\bar{n} \ll \frac{1}{2}$) which applies to these monomers. The insolubility of the polymer in vinyl chloride does not seem to be of major importance, perhaps because the polymer is swollen by the monomer to a surprisingly large extent.[12] It has recently been observed[13] that monomers of which the solubility in water is even lower than that of styrene (e.g. dodecyl methacrylate, *t*-butyl styrene) do not undergo emulsion polymerization when a water soluble initiator is used unless acetone or methanol is added to increase their solubility in the aqueous phase.

NUCLEATION OF LATEX PARTICLES BY OLIGOMERIC PRECIPITATION

Despite the prominence given to the micellar nucleation of latex particles in the Harkins[2] theory, and although this is certainly the principal particle nucleation process in the polymerization of styrene in the presence of micellar emulsifier, neither the presence of micelles nor even of a surfactant below its critical micelle concentration is an essential feature of emulsion polymerization, provided a stable latex can be formed. Electrostatic stabilization of a poly(vinyl acetate) latex by ionic end-groups derived from the initiator was first observed by Priest in 1952.[14] The essential characteristic of emulsion polymerization as emphasized by Haward in 1949[15] is the *isolation of individual polymerizing radicals in discrete loci*.

Persulphate is the most widely used initiator for emulsion polymerization reactions. It provides ionic end-groups capable of stabilizing latex particles in the absence of surfactants, although sulphate end-groups are liable to acid catalysed hydrolysis giving hydroxyl end-groups which may possibly be oxidized to carboxyl by the persulphate.

The outcome of polymerizations of aqueous solutions and emulsions of vinyl acetate depends vitally on the initiator selected. So far as solutions are concerned, although vinyl acetate is relatively soluble in water, poly(vinyl acetate) is not, and precipitates as soon as it is formed. Monomer is then partitioned between the phases according to a curious relationship[16]

$$(M_p) = 13{\cdot}7(M_w)^2$$

where (M_p) and (M_w) are the monomer concentrations by weight in the polymer and water phases respectively. The form of this relation has been confirmed independently:[17] it arises because the miscibility of vinyl acetate with water is limited whereas vinyl acetate and poly(vinyl acetate) are miscible in all proportions, and consequently the activity of the monomer increases much more rapidly than its concentration in the aqueous phase. The H_2O_2/Fe^{3+} system does not introduce charged end-groups capable of stabilizing electrostatically the poly(vinyl acetate) latex which is produced. Consequently the polymer particles coalesce and (in the absence of stirring) sink to the bottom of the reaction vessel extracting most of the monomer which remains. Because of the low interfacial area, radicals generated in the water phase are unable to reach the remaining monomer in the polymer phase so that the polymerization slows down and stops without all the monomer being polymerized. In the experiments shown in Fig. 2,[16] the concentrations of persulphate and H_2O_2/Fe^{3+} were chosen to give the same rate of initiation,

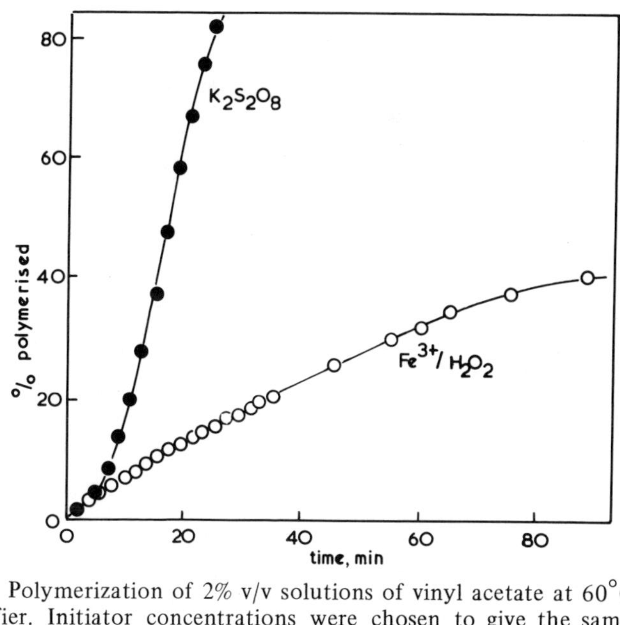

Fig. 2. Polymerization of 2% v/v solutions of vinyl acetate at 60°C without emulsifier. Initiator concentrations were chosen to give the same rates of radical production.[16]

and it can be seen that the initial rates of polymerization are actually identical; but when persulphate was used and a stable latex was produced, the rate of polymerization soon accelerated by a factor of about twenty and the conditions are those of an emulsion polymerization during Interval III when monomer droplets have disappeared and virtually all the remaining monomer is in the latex particles. These observations have recently been confirmed by Hayashi and Hoje.[18]

EFFECT OF IONIC STRENGTH

The ionic strength of the aqueous phase affects the stability of the latex particles and the rate of reaction. The effect of increasing ionic strength is to reduce the electrostatic repulsive energy barrier so that the latex ultimately coagulates. When emulsifier is present this factor is less obvious and it has generally been neglected, although it is taken into account by Klein *et al.*[19] and by Goodwin *et al.*,[20] cf. Fig. 3.

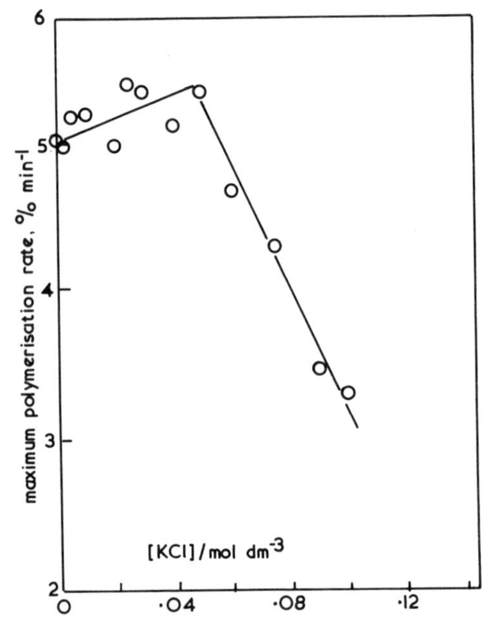

Fig. 3. Effect of increasing ionic strength on the rate of polymerization of 2% v/v vinyl acetate solutions initiated by 0·02% $K_2S_2O_8$ at 60°C.[22]

EFFECT OF OXYGEN

Although oxygen inhibits the polymerization of vinyl acetate, it does not have a critical effect on the course of the reaction (Fig. 4). The course of the reaction when a little oxygen is present initially appears to be identical after the end of the inhibition period with that observed when oxygen is removed completely by vacuum degassing.[16] Nevertheless, soluble oligomer is formed during the inhibition period which is surface active and which is adsorbed on the latex particles once they form, reducing their size slightly compared with that observed in the complete absence of oxygen. With a 0·232 mol dm^{-3} solution of vinyl acetate with $7{\cdot}4 \times 10^{-4}$ mol dm^{-3} persulphate, the diameter of a particle with the weight-average mass, $\bar{d}_T$, was found by a light scattering method[21] to be 293 nm when vacuum degassing was used, 260 nm when most of the oxygen was displaced with a stream of nitrogen, and 219 nm when the oxygen concentration was increased by passing in oxygen.[22]

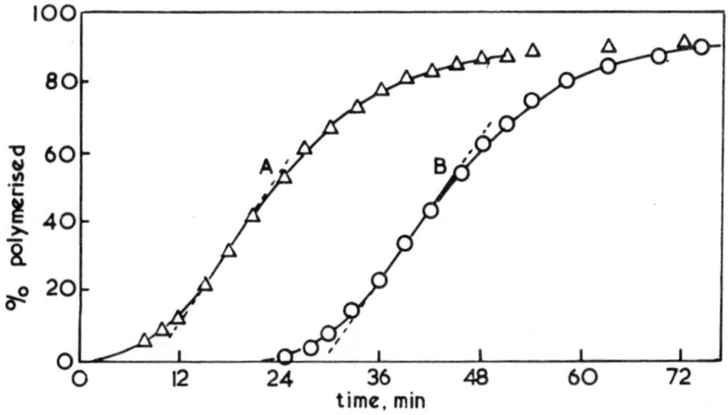

Fig. 4. Effect of oxygen on polymerization of 1% v/v aqueous solutions of vinyl acetate at 60°C with 0·01% $K_2S_2O_8$.[16] (A) Rigorously degassed: maximum rate 3·24% min^{-1}. (B) Without degassing: maximum rate 3·18% min^{-1}.

Morris and Parts[23] found that the *rate of decomposition* of persulphate was greatly accelerated in the presence of oxidizable substrates (including, notably, vinyl acetate) by induced chain decomposition:

$$S_2O_8^{2-} \rightarrow 2SO_4^{-}\cdot$$

$$SO_4^{-}\cdot + RH \rightarrow HSO_4^{-} + R\cdot$$

$$R\cdot + S_2O_8^{2-} \rightarrow SO_4^{-}\cdot + SO_4^{2-} + R$$

Induced chain decomposition does not necessarily increase the *rate of radical formation* and is, in any case, likely to be less important at the concentrations of persulphate used to initiate polymerization, which are much lower than those required for titrimetric determinations of persulphate concentration according to the procedure used by Morris and Parts. The rate of consumption of the inhibitor diphenylpicrylhydrazyl was used to measure the rate of radical formation at the concentrations of persulphate used to initiate polymerization.[24] The initial rate of consumption of the inhibitor (Fig. 5) is precisely that expected on the basis of Kolthoff and Miller's rate constant for persulphate decomposition,[25] which is the value which is most widely used although there are many other higher values to be found in the literature. However, the rate of consumption of diphenylpicrylhydrazyl seems to be autocatalytic. Nevertheless, it appears that under conditions in which radicals can initiate polymerization as soon as they are formed no acceleration of the decomposition rate will be observed. Although the arsenious oxide-bromate titration used by Morris and Parts[23] is suitable for the determination of persulphate concentrations in the absence of additives, it was not successful in the presence of oxidizable substrates in our experiments.

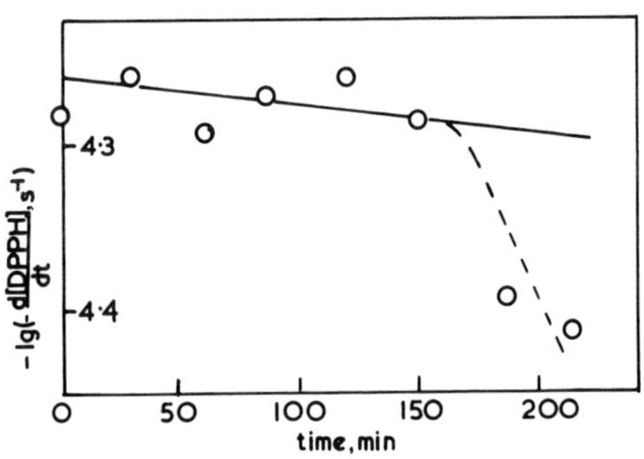

Fig. 5. Rate of decomposition of potassium persulphate in presence of vinyl acetate at 60°C.[24] Persulphate concentrations were measured by determining the rate of consumption of diphenylpicrylhydrazyl. The full line indicates the concentrations expected on the basis of Kolthoff and Miller's decomposition rate constant.[25]

AZONITRILE SULPHONATE INITIATORS

Although persulphate is unlikely ever to be displaced from its pre-eminence as the initiator for emulsion polymerization on an industrial scale, it is a peroxide and is therefore not an ideal initiator for kinetic work. Even if induced chain decomposition and acid catalysed decomposition do not increase the rate of radical formation they do increase the rate of decomposition of the initiator so that the rate of initiation may decrease more rapidly than would be expected. The introduction of the azonitrile initiators which are not subject to induced chain decomposition brought a great simplification to bulk polymerization kinetics. Sulphonated azonitrile initiators were claimed by Rhone-Poulenc[26] in 1960 as a means of introducing anionic end-groups into polyacrylonitrile fibre to render it dyeable with basic dyes. Better methods of achieving this object have subsequently been developed but these initiators (I, II)

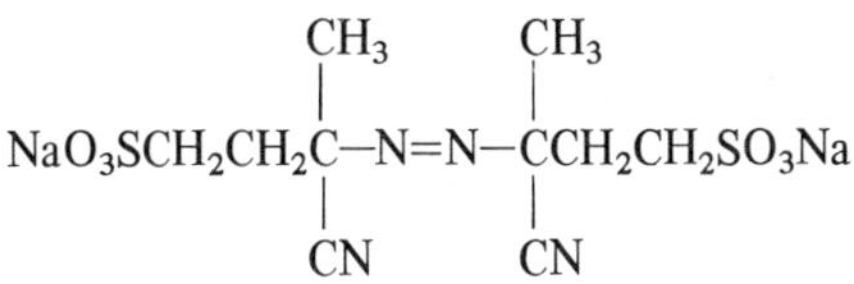

(I)

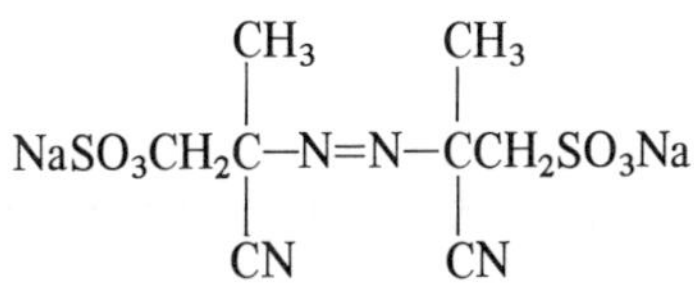

(II)

would be useful in studies of emulsion polymerization kinetics and as a means of introducing strong acid end-groups which would not be liable to acid-catalysed hydrolysis into polymer colloids. Although the preparation of the corresponding initiator with carboxyl end-groups (bis-α-azovaleronitrile[27]) is not difficult, the Strecker synthesis for the azonitrilesulphonates has proved to be very difficult and the yields which have been obtained[28] are poor. A reliable preparation for an azonitrilesulphonate would be very welcome.

DLVO THEORY

Electrostatically stabilized polymer latex particles are of the nature of a hydrophobic colloid, and consequently the theory of the stability of lyophobic colloids developed by Derjaguin and Landau and by Verwey and Overbeek[29] can be applied to them. Primary latex particles precipitating from aqueous solution are not stable, and coalesce until the surface charge density of the particles has increased sufficiently to counteract the van der Waals attractive forces reducing the rate of coalescence to zero. Unfortunately, the relations between particle size and the surface charge density, surfactant concentration, and ionic strength required for stability are complicated, so that there is no simple criterion for the size which the final particles in an emulsion polymerization will attain. One simple principle did however emerge from our calculations on this subject.

When enough emulsifier is present it will stabilize small particles by adsorption, but, in absence of emulsifier small particles are less stable than large particles with which they consequently coalesce. This is why a monodisperse particle size distribution is obtained when particles are nucleated by oligomeric precipitation in absence of emulsifier. New particles continue to be formed throughout the reaction but they are unstable and, once a sufficiently high concentration of pre-existing particles has been built up, they coalesce with these rather than with each other. Thus the number of particles soon becomes constant.

EFFECT OF VINYL ACETATE ON THE DENSITY OF WATER

A puzzling feature of the polymerization of aqueous solutions of vinyl acetate is that the percentage contraction corresponding to 100% polymerization is 15·7% at 60°C[16] where the value calculated from the densities of the monomer and the polymer is 23·6%, which is the value which is found in bulk polymerization.[30] This is a result of the effect of the solute on the structure of water which even at 60°C retains a large proportion of the tetrahedral hydrogen-bonded structure of ice. Hydration of vinyl acetate increases the mean density of water (Table I[22]). Poly(vinyl acetate) latex has no effect on the density of water: presumably it is not hydrated to a significant extent. Consequently, when a solution of vinyl acetate polymerizes the water expands, partially compensating the contraction due to polymerization. When a separate monomer phase is present, intermediate values of the percentage contraction are applicable, which makes it difficult to apply the dilatometric method of determining polymerization rates to emulsions of vinyl acetate.

TABLE I
Effect of vinyl acetate on the density of water

Solution	Density (g cm^{-3}) at 60°C		
	of solution	*of solute*	*of solvent*
Water	0·98324		
0·02% $K_2S_2O_8$ in water	0·9832		
1% v/v Vinyl acetate	0·9832 ± 0·0002	0·9811	0·9855
2% v/v Vinyl acetate	0·9832 ± 0·0001	0·9810	0·9854
Vinyl acetate	0·8796 ± 0·0003		
1% Water in vinyl acetate	0·8808 ± 0·0001	0·8807	
1·76% Poly(vinyl acetate) suspension	0·9866 ± 0·0001	0·9865	
Poly(vinyl acetate)	1·162		

Contraction of vinyl acetate on 100% polymerization at 60°C: in bulk 23·6%, in water 15·7%.
From reference 22.

EFFECT OF POLYVINYL ALCOHOL ON POLYMERIZATION OF VINYL ACETATE

Polyvinyl alcohol is a common constituent of recipes for the emulsion polymerization of vinyl acetate. Its principal function is as a thickener for the aqueous phase: alternative water soluble polymers such as ethyl cellulose and hydroxymethyl cellulose of appropriate degrees of substitution may be used for this purpose.

However, polyvinyl alcohol has other effects apart from improving the rheology of the latex. The type of commercial polyvinyl alcohol which is preferred for this application is that with about 12 mole % residual acetyl groups produced by alkaline methanolysis of poly(vinyl acetate). The residual acetyl groups are not randomly distributed but occur in short blocks.[31] These materials are polymeric surfactants. In most practical recipes a minor proportion of an anionic surfactant–often sodium dodecyl sulphate–is also used. The anionic surfactant is much more effective than polyvinyl alcohol in reducing the size of the poly(vinyl acetate) latex particles formed but it has little effect on the rate of the reaction.[16] By contrast, the sample of polyvinyl alcohol originally used (DuPont's 'Elvanol' 52-22), which was selected as being typical of the material in commercial use at the time, reduces the rate of polymerization of vinyl acetate solutions to about half its original value even at a concentration of only 0·36% (much larger concentrations, e.g. 5%, are used in emulsion polymerization recipes). The effect of polyvinyl alcohol depends also on monomer concentration: the rate of polymerization of vinyl acetate emulsions does increase with concentration of the same polyvinyl alcohol. The work of Donescu and co-workers[32] has done much to elucidate this effect of the amount of monomer: under some conditions considerable amounts of graft copolymers are formed.

Later we wished to vary the viscosity of the aqueous phase and obtained for this purpose a series of 'Gohsenol' (Nippon Gohsei, Japan) polyvinyl alcohols of the same acetyl content but differing degrees of polymerization. Surprisingly, the effect of these on the rate of polymerization of vinyl acetate solutions was entirely different: any retardation of the rate of polymerization was slight.

By contrast, other samples, e.g. 'Polyviols' (Wacker Chemie, West Germany), had effects which were as great as that of the original sample. We looked at various properties of the samples without avail. None of them seemed to correlate with the observed retardatory effect. The samples did seem to fall into two groups:[33] those which were strongly adsorbed on poly(vinyl acetate) latex, and those which were comparatively weakly adsorbed; but the adsorption isotherms did not correlate with the retardatory effect. Subsequently Fleer *et al.*[34] showed that the adsorption of polyvinyl alcohol is essentially irreversible. Consequently it is vital that the polyvinyl alcohol solution should be diluted before being mixed with the latex. The apparent strong adsorption probably resulted from mixing the latex and the polyvinyl alcohol stock solution before diluting to the final volume. When the correct procedure is followed, the weak types of isotherms are obtained,[35] although these do differ between samples.

THE PROBLEM OF THE VARIABILITY OF POLYVINYL ALCOHOLS

The problem of variability in the properties of polyvinyl alcohols has evidently also been encountered in the Soviet Union[36] where it was suggested that a condensation product of acetaldehyde formed by hydrolysis of residual vinyl acetate is responsible for this effect. The condensation product could be extracted with methanol, and after extraction retardation of the rate of polymerization by polyvinyl alcohol was no longer observed. However, this work appears to relate to experimental samples which had been deliberately underpolymerized: residual monomer levels as high as 6% would probably not be encountered in normal industrial practice. The acetaldehyde condensation product was identified by its ultraviolet absorption spectrum. Slight heating of polyvinyl alcohol increases its ultraviolet absorption as a result of elimination of water and slight oxidation resulting in a low concentration of sequences of one, two, or three double bonds conjugated to carbonyl groups in the polyvinyl alcohol chain:[37] enhanced conjugated carbonyl content could easily result from the temperatures encountered when the polymer powder is dried.

There is a correlation between the intensity of absorbance of polyvinyl alcohols at either 225 nm

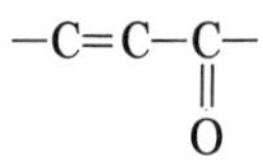

or 285 nm

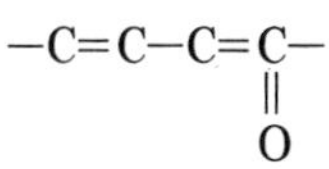

and the retardation of the rate of polymerization of vinyl acetate solutions, although the coefficient of correlation is less than 1·0 indicating that some other factor may also contribute to this effect. Heating samples to only 100°C for 10 min increases the ultraviolet absorption and the retardatory effect. Extraction of commercial samples with methanol does reduce the ultraviolet absorption and retardatory effect of some samples but does not eliminate either. Hence, although any residual vinyl acetate can, on hydrolysis, form condensation products which retard polymerization, most of the conjugated unsaturation which causes this effect is in the polymer itself and cannot be extracted with methanol. It is not clear how this unsaturation arises. It is not necessarily deleterious: Noro[38] varied the carbonyl content of polyvinyl alcohol by preparing it from the copolymer obtained from vinyl acetate with minor proportions of acrolein and found that increasing carbonyl content of the polyvinyl alcohol improved the stability of a poly(vinyl acetate) latex prepared with it to repeated freezing and thawing, and that variation of carbonyl content of a completely hydrolysed polyvinyl alcohol by a factor of four resulted in a nine-fold increase in the viscosity of the polyvinyl alcohol latex produced.

EFFECT OF POLYVINYL ALCOHOL STRUCTURE ON POLY(VINYL ACETATE) LATEX VISCOSITY

The viscosity of polyvinyl alcohol latexes is very sensitive to the properties of the polyvinyl alcohol used in their preparation. This may be partly a function of the average particle size and the particle size distribution of the latex formed: increased adsorption of polyvinyl alcohol will result in a reduction of average particle size. However, Shiraishi[39] has shown that latex viscosities may differ markedly even when the average particle size of the latex is the same (although it is possible that the particle size distributions might differ). We examined two polyvinyl alcohols which conformed to the same specification —acetyl content 12 ± 1 mole %, aqueous solution viscosity at a concentration of 4%, 30 ± 5 cP. One of these was a standard grade of Japanese manufacture, 'Gohsenol' GH-17, the other, sample B, was of undisclosed origin. The actual 4% solution viscosities were respectively 26·4 and 27·5 cP but when B was directly substituted for GH-17 in a standard emulsion polymerization recipe the latex viscosity obtained was only 8·9 P instead of 16·3 P. Reformulation with an increased quantity of B would be required to restore the original latex viscosity. The samples appeared to be identical in most respects—X-ray powder diffraction, infrared spectra, ^{13}C nuclear magnetic resonance chemical shifts. The crystalline melting point of polyvinyl alcohol is reduced by copolymerization with vinyl acetate in a way which depends on the distribution of the acetate groups[31] and a slight difference in melting point was observed by Differential Thermal Analysis (DTA). However, the DTA melting point is very sensitive to the precise acetyl content which differed by 1 mole % between the samples. When this was taken into account, the derived average acetyl sequence length, $\bar{n} = 2{\cdot}7$, was identical. The ultraviolet absorbance of the samples was not identical, being somewhat lower for sample B. But the difference in the conjugated monoene carbonyl content (GH-17 0·67 mole %, B 0·50 mole %) seems to be too small to account for the observed difference in latex viscosity.

The main features of the ^{13}C NMR spectrum of polyvinyl alcohol are the methylene (CH_2) and methine (CH) bands. A methyl (CH_3) band due to the residual acetyl groups may also be observed. The methine band is split first into components attributable to CHOH and CHOAc methines each of which, secondarily, shows fine structure attributable to groups in isotactic, heterotactic, and syndiotactic placements. The methylene band has three main components attributable to a methylene group (a) between two hydroxyl groups, (b) between a hydroxyl group and an acetyl group, and (c) between two acetyl groups; each of these has fine structure due to tacticity. From the

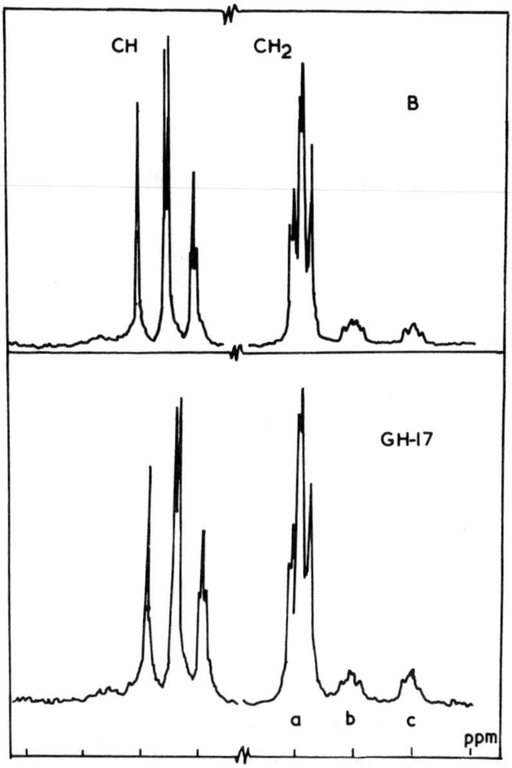

Fig. 6. ^{13}C NMR spectra of two polyvinyl alcohols containing approximately 12 mole % residual acetyl groups obtained under identical conditions under which quantitative intensities are obtained despite the use of proton decoupling. The intensities of the sum of the methylene peaks is lower relative to that of the methine peak for B than for GH-17. The splitting of the main peaks is attributable to the different possible steric placements of consecutive groups. The methylene peak (a) is attributable to the structure —CHOH—$\underline{C}H_2$—CHOH, (b) to —CHOH—$\underline{C}H_2$—CHOAc—, and (c) to —CHOAc—$\underline{C}H_2$—CHOAc.

integrations (Fig. 6) it is apparent that the ratio $\Sigma CH/\Sigma CH_2$ differs slightly between the samples when the spectra are run under identical conditions. Unfortunately, the conditions used do not seem to have been quite the optimum for quantitative determination because the ratio obtained with a sample prepared by photosensitized polymerization at $-25°C$, which was expected to be genuinely linear[40] under the same operating conditions, was

0·95. However, the ratio obtained for the linear sample could be raised to 0·99 by modifying the operating conditions. The ratios for GH-17 and B differed from each other and were both lower than for the linear sample. In general, secondary hydrogen atoms have a lower bond dissociation energy than primary hydrogen atoms and are therefore more reactive in transfer reactions. Transfer to the secondary hydrogen atom of the methine group transforms it into a quaternary carbon atom which has a long relaxation time and is difficult to detect in low concentration by ^{13}C NMR spectroscopy. This reduces the CH/CH_2 ratio. The concentration of branch points is probably about 2 mole % in GH-17 and 1 mole % in B. These must presumably be mostly short-chain non-hydrolyzable branches formed by intramolecular transfer because the frequency of long-chain non-hydrolyzable branches formed by intermolecular transfer to the main chain is believed to be considerably lower than this. The adsorption isotherms of the samples on emulsifier-free poly(vinyl acetate) latex differ considerably (Fig. 7[35]), GH-17 being more strongly adsorbed. GH-17 is also more strongly adsorbed at the air/water interface (Fig. 8). The shallower curvature of the surface tension–concentration

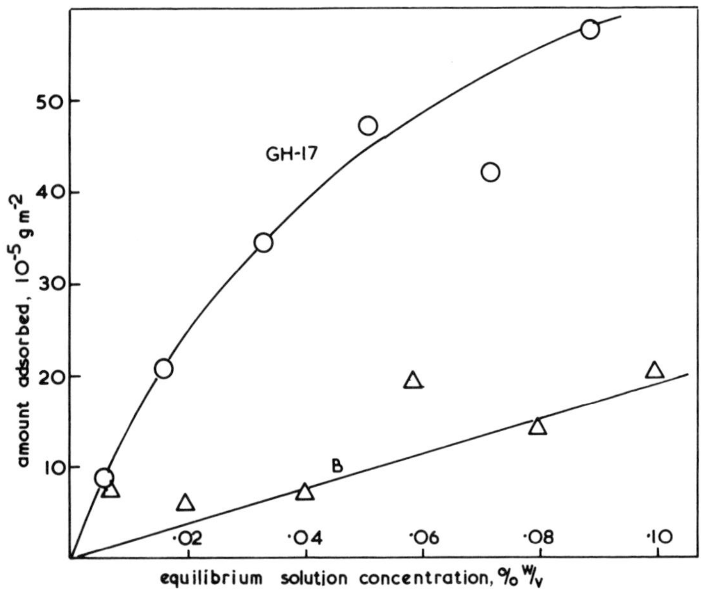

Fig. 7. Adsorption isotherms of polyvinyl alcohols on emulsifier-free poly-(vinyl acetate) latex at approximately 20°C.[35]

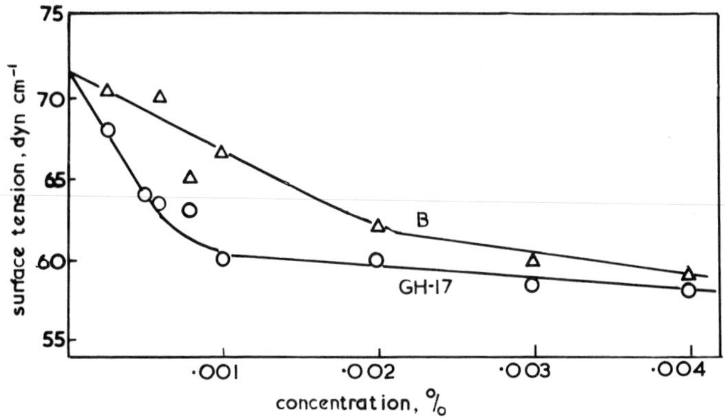

Fig. 8. Lowering of surface tension of water by polyvinyl alcohols at approximately 20°C.[41]

graph for B suggests that it has a broader distribution of acetyl sequence lengths despite the average acetyl sequence length being the same for the two samples.

CONCLUSION

Despite the fact that the water solubility of vinyl acetate is a major factor which causes the characteristics of its polymerization in emulsion to differ from those of styrene, the study of the polymerization of aqueous solutions of vinyl acetate in the absence of a separate monomer phase has not really contributed much to the elucidation of these differences because these polymerizations have essentially the character of an emulsion polymerization in Interval III after all the monomer has been absorbed by the polymer phase. Nevertheless, some useful observations have been made. Stable, monodisperse latexes can easily be produced without the use of an emulsifier which could be stabilized by a single type of well characterized ionic end-group if a suitable initiator system were used. It has been shown that the only important particle nucleation mechanism when the monomer is sufficiently soluble in water is oligomeric precipitation, when the effect of solubilization of monomer in emulsifier micelles becomes irrelevant. Conventional emulsifiers are much less well adsorbed by polar polymers than by non-polar polymers but polyvinyl alcohol (with 12 mole % residual acetyl) is compatible with poly(vinyl acetate) and its use is often required to obtain a sufficiently stable latex for many applications. Unfortunately, small differences in the structure of the polyvinyl alcohol used have a profound effect on the properties of the latex obtained. The identification of the nature and origin of these structural differences is difficult but there is evidence that small differences in the extent of short main-chain branching have a large effect on the surface activity of polyvinyl alcohols.

ACKNOWLEDGEMENTS

The results reviewed here were mostly obtained by Drs P. A. Taylor, L. C.-H. Kwan (née Chong), C. J. Tonge and S. R. Naravane whose work was made possible by Research Scholarships provided by Vinyl Products Ltd, the Science Research Council, and Air Products and Chemicals Inc. Research projects undertaken by UMIST undergraduates have also made useful contributions.

REFERENCES

1. F. A. Bovey, I. M. Kolthoff, A. I. Medalia and E. J. Meehan, *Emulsion Polymerization,* New York, Interscience, 1955.
2. W. D. Harkins, *J. Chem. Phys.,* **13**, 381 (1945); *J. Amer. Chem. Soc.,* **69**, 1428 (1947); *J. Polym. Sci.,* **5**, 217 (1950).
3. W. V. Smith and R. H. Ewart, *J. Chem. Phys.,* **16**, 592 (1948).
4. E. Bartholome, H. Gerrens, H. Herbeck and R. M. Weitz, *Z. Elektrochem.,* **60**, 334 (1956).
5. G. M. Burnett and R. S. Lehrle, *Proc. Roy. Soc. (London),* **A253**, 331 (1959).
6. J. H. Baxendale, M. G. Evans and G. S. Park, *Trans. Faraday Soc.,* **42**, 155 (1946).
7. F. S. Dainton and P. H. Seaman, *J. Polym. Sci.,* **39**, 279 (1959).
8. S. Okamura and T. Motoyama, *J. Polym. Sci.,* **58**, 221 (1962).
9. J. L. Gardon, *J. Polym. Sci., A-1,* **6**, 643 (1968).
10. J. Ugelstad, P. C. Mørk and J. O. Aasen, *J. Polym. Sci., A-1,* **5**, 2281 (1967).
11. K. H. S. Chang, M. H. Litt and M. Nomura, this volume, Chapt. 6.
12. A. R. Berens, *Angew. Makromol. Chem.,* **47**, 97 (1975).
13. F. W. Waite, ICI Ltd (Paints Division), Slough, private communication.
14. W. J. Priest, *J. Phys. Chem.,* **56**, 1077 (1952).
15. R. N. Haward, *J. Polym. Sci.,* **4**, 273 (1949).
16. A. S. Dunn and P. A. Taylor, *Makromol. Chem.,* **83**, 207 (1965).
17. D. H. Napper and A. G. Parts, *J. Polym. Sci.,* **61**, 113 (1962).
18. S. Hayashi and N. Hoje, *Makromol. Chem.,* **177**, 1215 (1976).

19. A. Klein, C. H. Kuist and V. T. Stannett, *J. Polym. Sci., Polym. Chem. Ed.*, **11**, 2111 (1973).
20. J. W. Goodwin, J. Hearn, C. C. Ho and R. H. Ottewill, *Br. Polym. J.*, **5**, 347 (1973).
21. G. M. Burnett, R. S. Lehrle, D. W. Ovenall and F. W. Peaker, *J. Polym. Sci.*, **29**, 417 (1958).
22. C. J. Tonge, Ph.D. Thesis, Manchester, 1972.
23. C. E. M. Morris and A. G. Parts, *Makromol. Chem.*, **119**, 212 (1968).
24. A. S. Dunn and C. J. Tonge, *Polymer Preprints*, **13**, 1261 (1972).
25. I. M. Kolthoff and I. K. Miller, *J. Amer. Chem. Soc.*, **73**, 3055 (1951).
26. J. A. Phelisse and C. A. Quiby (Rhone-Poulenc), US Patent 3161630, 1964.
27. C. H. Bamford, A. D. Jenkins and R. P. Wayne, *Trans. Faraday Soc.*, **56**, 930 (1960).
28. L.-J. Liu and I. M. Krieger, in *Emulsions, Latices, and Dispersions*, P. Becher and M. N. Yudenfreund (eds.), New York, Dekker, 1979, p. 344.
29. (a) B. V. Derjaguin and J. D. Landau, *Acta phys.-chim. USSR*, **14**, 633 (1941).
(b) E. J. W. Verwey and J. Th. G. Overbeek, *Theory of the Stability of Lyophobic Colloids*, Elsevier, Amsterdam, 1948.
30. H. W. Starkweather and G. B. Taylor, *J. Amer. Chem. Soc.*, **52**, 4708 (1930).
31. R. K. Tubbs, *J. Polym. Sci., A-1*, **4**, 623 (1966).
32. V. Dimonie, D. Donescu, M. Munteanu, C. Hagiopol and I. Gavat, *Rev. Rouman. Chim.*, **19**, 903 (1974).
33. A. S. Dunn, C. J. Tonge and S. A. B. Anabtawi, in *Emulsion Polymerization*, I. Piirma and J. L. Gardon (eds.), Washington, ACS Symposium Series No. 24, 1976, p. 24.
34. G. J. Fleer, L. K. Koopal and J. Lyklema, *Koll. Z. Z. Polym.*, **250**, 689 (1972).
35. R. B. Abdul Ghani, unpublished work; cf. A. S. Dunn, *Chem. Ind.* (in press).
36. V. T. Shirinyan, S. S. Mnatsakanov, G. A. Shirikova, F. O. Poznyakova, G. S. Popova and T. V. Khoostyntseva, *Plast. Massy*, No. 2, 15 (1974); English translation: *Int. Polym. Sci. & Techn.*, **2**, 82 (1975).
37. B. Duncalf and A. S. Dunn, *J. Polym. Sci., C*, **16**, 1167 (1967).
38. K. Noro, *Br. Polym. J.*, **2**, 128 (1970).
39. M. Shiraishi, *Br. Polym. J.*, **2**, 135 (1970).
40. G. M. Burnett, M. H. George and H. W. Melville, *J. Polym. Sci.*, **16**, 31 (1955).
41. D. K. Powell, unpublished work.

3

Effects of Particle Number and Initiator Level on the Kinetics of Vinyl Acetate Emulsion Polymerizations

RICHARD L. ZOLLARS†
Department of Chemical Engineering, Washington State University, USA

SUMMARY

The kinetics of the emulsion polymerization of vinyl acetate have been studied in numerous previous investigations. These previous investigations agree on a number of phenomena, namely that the emulsion polymerization of vinyl acetate does not follow Smith-Ewart kinetics and that the reaction exhibits a constant reaction rate well beyond the point at which the separate monomer phase disappears. These phenomena are thought to occur because of the high water solubility of vinyl acetate and the high rate of transfer to monomer. On the other hand, much of the information presented by previous investigators appears to be contradictory, and none of the previously suggested models for the reaction can account for all of the phenomena observed.

The results to be presented indicate that the variability observed by previous investigators is not a consequence of variations in experimental materials or techniques. Instead the variability appears to be a result of changes in the behavior of the particle number with changes in the ionic strength of the aqueous phase. Using the results of previous investigations and the results of experiments carried on for this investigation, a reaction model is proposed which is consistent with all previously reported data.

EXPERIMENTAL

Vinyl acetate monomer, freed of all inhibitors, was obtained from the South Charleston, West Virginia plant of the Union Carbide Corporation. Deionized

† Work performed while author employed at the Union Carbide Technical Center, South Charleston, West Virginia 25303, USA.

water, normally supplied to the laboratories was used without further purification. Potassium persulfate, used as the initiator, and potassium sulfate, used to adjust ionic strength, were reagent grade chemicals obtained from Merck, Inc. and used without further treatment. The emulsifier, sodium lauryl sulfate, was obtained as a 99% pure powder from Alcolac, Inc. (Sipex WD crystals). A similarly pure grade of the other emulsifier used, sodium cetyl sulfate, could not be obtained. Instead, the purified emulsifier was obtained from Sipex EC-111 (Alcolac, Inc.) by a process described in a previous report.[1]

A uniform procedure was used for all experiments performed during this investigation to ensure that any variation did not occur because of changes in experimental procedure. Water (1817 g) plus all emulsifiers and neutral salts were added to the reactor and the mixture stirred vigorously to ensure complete dispersion of the emulsifier. The vinyl acetate then was added (614 g) and the system was thoroughly purged with high purity nitrogen while the reactor temperature was adjusted to the desired level. During this time the desired amount of the initiator was dissolved in 25 g of water and heated to the reaction temperature. After the reactor temperature had stabilized (about 30 minutes) the contents were sampled through a septum in the reactor cover using a syringe, and the initiator solution was added. After a brief induction period the polymerization reaction started. Throughout the reaction a number of samples were removed from the reactor, again using a syringe so that the nitrogen purge would not be broken. The samples were short-stopped with hydroquinone for later analysis.

The samples that had been withdrawn were analyzed for monomer composition using a gas chromatograph. From these results a conversion versus time history for the reaction was determined and used to obtain the reaction rate during the constant rate period. Particle size measurements were obtained using a dissymmetry light scattering method. This technique yields an average particle diameter which is approximately equal to the weight average particle diameter. To convert the particle number obtained from this measurement to the number average particle number required for kinetic calculations, a polydispersity of 2·6, observed by other investigators,[2,3] was used.

RESULTS

The results from a number of other investigations of the emulsion polymerization of vinyl acetate using potassium persulfate as the initiator and sodium lauryl sulfate as the emulsifier are shown in Table I. While all of these investigations report similar general behavior, their differences can best be noted by observing the different behavior observed for both the reaction rate order and particle number order with respect to the initiator level. Among the investigations reported in Table I there appears to be a split amongst the investigations into two different viewpoints. There is a group of investigations which observed that the overall reaction order with respect to the initiator level was half order while reporting that there appeared to be no dependence of the particle number on the initiator level. There also appears to be another group which observed a first order dependence for both the particle number and overall reaction rate on the initiator concentration. Furthermore, those investigators reporting the latter behavior appear to be those that worked with a reaction temperature of 60°C while the former behavior was reported mainly by investigators working at a 50°C reaction temperature.

TABLE I
Results of previous investigations of vinyl acetate emulsion polymerizations

Investigators	*Emulsifier*	*$K_2S_2O_8$ range ($M \times 10^3$ M)*	*Temp. (°C)*	*Reaction rate order on initiator*	*Particle number order on initiator*
1. Dunn and Taylor[12]	SLS	0·185–1·48	60	0·64	0·35
2. Patsiga *et al.*[4]	SLS	0·8	60	0·8–1·0†	—
3. Priest[3]	SLS	0·12–0·16	50	—	0·0
4. Dunn and Chong[10]	SLS	0·1–0·8	60	0·6–0·9	0·6–0·9
5. Nomura *et al.*[13]	SLS	4·62	50	0·5	0·0
6. Litt *et al.*[5]	SLS	0·25–2·0	60	1·0	1·2
7. Gershberg[14]	SLS	1·0–30·0	50	0·6	—
8. Friis and Nyhagen[2]	SLS	0·5–4·0	50	0·5	0·0

† Seeded polymerizations.

In addition to the experimental results presented in Table I two recent papers have proposed detailed reaction models for the emulsion polymerization reaction. Fortunately, one of these models, hypothesizing a polymer phase termination mechanism[2] is supported by the results at 50°C, while the other model, hypothesizing an aqueous phase termination mechanism[4–6] is supported by the results at 60°C. Thus, each of the two extremes of behavior for the reaction is described by a well-developed reaction model. The results from the present investigation will, therefore, be compared only with these two previous investigations as representatives of their respective groups.

The experiments performed during this investigation were designed to reproduce the temperature, monomer level, initiator level and ionic strength conditions used by the previous investigators so as to rule out the possibility that the previous differences in the results could have been caused by differences in experimental technique. Again, as with all of the investigations reported above, some common phenomena were observed. These phenomena are the disappearance of the separate monomer phase at about 20% conversion, the constant reaction rate between 20% and 80% conversion and the fact that the emulsifier level had little if any effect on the reaction rate.

Polymerization rate results

A series of polymerizations were carried out at varying initiator levels at both 50°C and 60°C, reproducing the conditions used by Friis and Litt, respectively. The results of the series of experiments at 50°C are shown in Fig. 1. Over the range of initiator concentrations used, 0·25 to $4{\cdot}0 \times 10^{-3}$ M, the reaction rate showed a $0{\cdot}72 \pm 0{\cdot}07$ power dependence on the initiator concentration. To

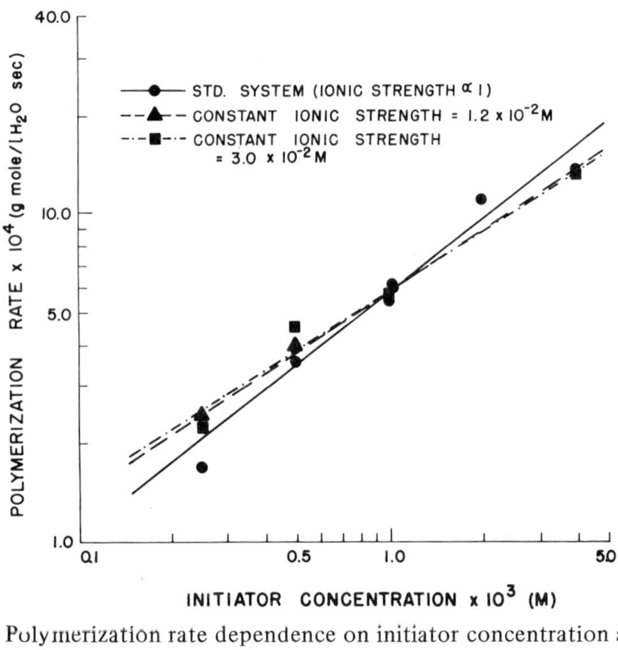

Fig. 1. Polymerization rate dependence on initiator concentration at 50°C.

compare these results with those of Friis, however, only the results at or above $1{\cdot}0 \times 10^{-3}$ M should be considered since all but one of the experiments by Friis were performed at this level or higher. In this range of initiator levels the results generated during this study indicate a reaction rate dependence of $0{\cdot}65 \pm 0{\cdot}07$ on the initiator concentration while Friis reports a $0{\cdot}56 \pm 0{\cdot}03$ power dependence. The results of a similar series of experiments carried out at 60°C are shown in Fig. 2. These results indicate a reaction rate dependence on the initiator level of $0{\cdot}88 \pm 0{\cdot}14$ while Litt reports the power to be $0{\cdot}93 \pm 0{\cdot}12$. Thus the results from both previous investigations are reproducible using a common technique and a common source of materials. The results presented in both Figs. 1 and 2 are the reaction rates observed at emulsifier levels of either 2·7 g/liter H_2O or 10·9 g/liter H_2O. No designation is made of the two emulsifier levels as there was no observable effect of emulsifier level on the results observed during this investigation. This agrees with the results reported by Litt and is also supported by the results presented by Friis, who reported only a $0{\cdot}12 \pm 0{\cdot}02$ power dependence of the reaction rate on the emulsifier level.

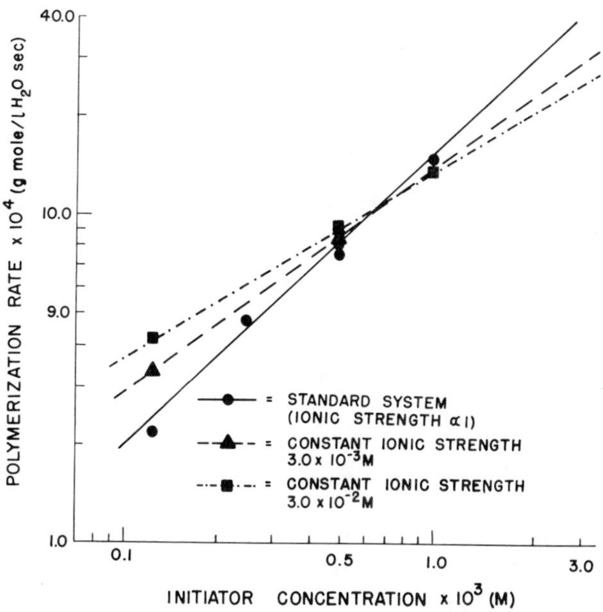

Fig. 2. Polymerization rate dependence on initiator concentration at 60°C.

The results reported above are those generated in what is called the standard system during this investigation. In the standard system the only salt added to the reaction mixture is the initiator, thus the ionic strength of the aqueous phase is controlled solely by the initiator level. Closely examining the results for this system in Fig. 1 indicates a trend towards a higher reaction rate order at the lower initiator levels. This trend was also observed by Friis, who reported that when his results at $0{\cdot}5 \times 10^{-3}$ M initiator were included the order of the reaction rose to 0·65. Friis attributed this increase in the order of the reaction to inhibition of the reaction by traces of oxygen in the gases used to blanket his reactor. Because of the higher temperatures used in the work by Litt, however, it is precisely in this region of initiator concentrations that previous investigators have reported the first order dependence of reaction rate on initiator concentration. The observed trend in the results at 50°C at these low initiator levels is towards a first order dependence.

The experiments performed using the standard system were then repeated, this time adding potassium sulfate in quantities sufficient to maintain a constant ionic strength at a level equivalent to the highest initiator concentration used. These results, at both 50°C and 60°C, are also shown in Figs. 1 and 2. Maintaining a constant ionic strength resulted in a decrease in the order of the reaction rate dependence on the initiator concentration to $0{\cdot}64 \pm 0{\cdot}09$ for 50°C and 0·68 for 60°C. A further increase in the ionic strength to a level equivalent to 0·01 M initiator caused no further change in the results at 50°C but reduced the order to 0·58 for the results at 60°C. Again, observing the changes occurring in the results only at initiator levels of $1{\cdot}0 \times 10^{-3}$ M initiator and higher at 50°C it can be seen that no significant changes have occurred. This is in agreement with the results presented by Friis in this range of initiator concentrations. The effects of the ionic strength on the results at 60°C are also in agreement with the observations of Litt.

Particle number results

At the same time that monomer analysis of the samples was being performed, each sample was also analyzed for particle size and, hence, particle number. Figures 3 and 4 represent the particle number behavior for emulsion polymerizations run under the standard conditions at 50°C and 60°C, respectively. Unlike reports from a number of previous investigators, the particle number observed during this investigation did not reach a constant level at 20% conversion, but continued to increase throughout the reaction. While the particle sizing technique used in this investigation is more sensitive to small particle generation than techniques such as electron microscopy, it has been used in other investigations, such as the investigation by Friis, where a constant

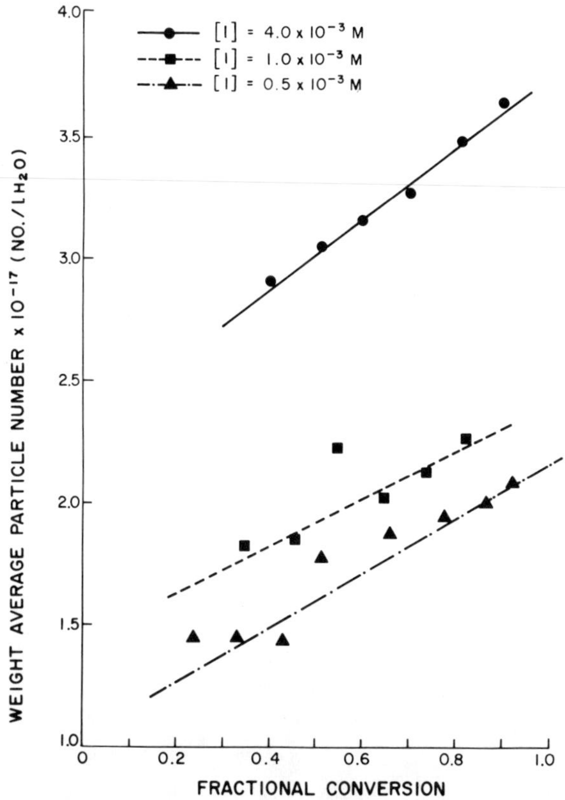

Fig. 3. Particle number dependence on conversion at 50°C.

particle number for conversions greater than 20% has been reported. The reasons for these differences are not apparent at this time.

Again, comparing the results from this investigation with those from the investigations by Friis and Litt, no differences were observed. In the range of initiator concentrations for which Friis reports particle number data, $0{\cdot}5 \times 10^{-3}$ to $1{\cdot}0 \times 10^{-3}$ M, Fig. 3 indicates that there is no significant change in the particle number with increasing initiator concentration. This behavior was also reported by Friis. Figure 4 indicates the same behavior at 60°C. However, when the initiator concentration is reduced to $0{\cdot}125 \times 10^{-3}$ M, the particle number decreases. This behavior was reported by Litt at these lower initiator concentrations. To determine the order of the dependence of the particle number on the initiation rate (or the initiator concentration for a constant temperature), the final particle number was plotted versus the rate of initiator

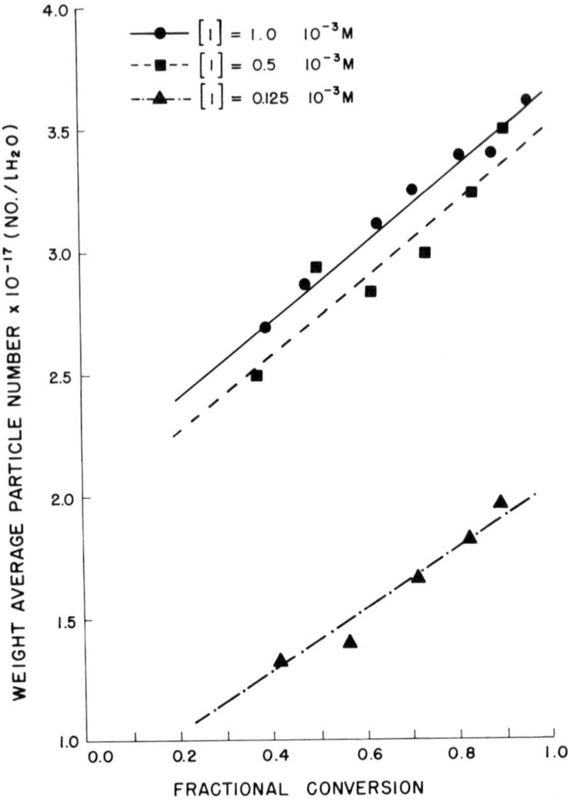

Fig. 4. Particle number dependence on conversion at 60°C.

decomposition, $2fk_i\,[I]$. The results, shown in Fig. 5, indicate a particle number which tends to increase with increasing initiator concentration, at either low initiator levels or high initiator levels. However, in the vicinity of an initiator concentration of $1{\cdot}0 \times 10^{-3}$ M, the particle number becomes relatively independent of the initiator concentration.

A number of the experiments performed above were repeated using sodium cetyl sulfate as the emulsifier rather than sodium lauryl sulfate. In a series of experiments performed using sodium alkyl sulfate surfactants of varying alkyl chain lengths, Napper and Alexander[7] reported that the polymerization rate and particle number both increased with increasing alkyl chain length. The results of the experiments performed during this investigation indicate no observable difference either in polymerization rate or particle

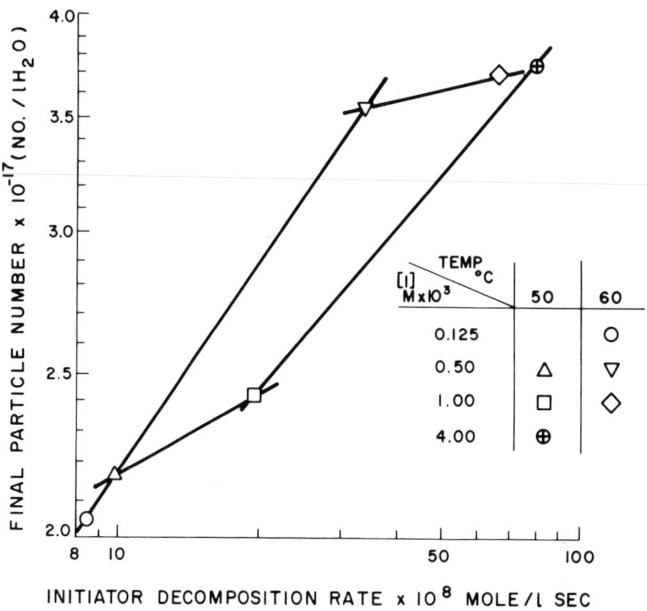

Fig. 5. Particle number dependence on initiator decomposition rate.

number between sodium lauryl or sodium cetyl sulfate. However, all of the experiments performed by Napper and Alexander were carried out at emulsifier concentrations less than the critical micelle concentration (CMC) and at monomer levels below the solubility of vinyl acetate in water, whereas the experiments in this investigation were carried out at emulsifier concentrations above the CMC and at a vinyl acetate concentration well above the solubility. The differences between the two investigations, therefore, may have arisen because of the higher concentrations used in this investigation.

One other experimental variable to be studied during this investigation was the intensity of the agitation. A pitched-blade turbine was used as the agitator for all experiments with the level of agitation being controlled by the speed of the agitator. Varying the level of agitation from a point where the agitation was just sufficient to prevent separation of the monomer phase up to a point where the reactor contents were being thrown up into the reflux condenser for the reactor produced no observable change in either reaction rate or particle number. Thus the hypothesis that variations in mixing intensity may have been the cause of the variable behavior of vinyl acetate emulsion polymerizations[6] was not substantiated.

DISCUSSION OF RESULTS

The comparison of the experimental data from this study with those of Friis and Nyhagen[2] or Litt *et al.*[5] was done not only because these two previous investigations were representative of the two extremes of the observed behavior, but also because these two sets of data were accompanied by well developed reaction models. Both the model proposed by Friis and that proposed by Litt have many points in common. Both models predict that the constant rate period observed during vinyl acetate emulsion polymerizations arises because of the high degree of transfer to monomer for vinyl acetate. After the disappearance of the separate monomer phase, the monomer concentration within each of the particles will decrease. This tends to decrease the reaction rate, but also decreases the rate of transfer to monomer. As the rate of transfer decreases, the rate of escape of free radicals from the particle also decreases, resulting in a higher radical concentration within the particles. The constant rate period arises because the higher radical concentration compensates for the decreasing monomer concentration. Friis, however, postulates that the termination reaction occurs in the polymer phase while Litt proposes that the termination occurs in the aqueous phase.

The model proposed by Litt has the following form:

$$R_p = C_1([I]\,[M_p]\,N_p)^{1/2}\left(1 + C_2\frac{[M_p]}{r^2[M_{aq}]}\right)^{-1/2} \tag{1}$$

where R_p is the rate of polymerization, $[I]$ is the initiator concentration (moles/liter H_2O), r is the radius of a particle, $[M_p]$ is the monomer concentration in the polymer phase (moles/liter), $[M_{aq}]$ is the monomer concentration in the aqueous phase (moles/liter), N_p is the particle number, and C_1 and C_2 are groups of rate constants. The group C_2 contains the rate constant for the sweep-up of aqueous phase radicals by the polymer particles, a constant which is dependent on the ionic strength of the aqueous phase.

The model proposed by Friis was a modification of the model originally proposed by Ugelstad *et al.*[8] for the emulsion polymerization of vinyl chloride and has the form

$$R_p = \frac{k_p[M_p]}{N_A}(2fk_i[I])^{1/2}\left(\frac{V_pN_A^2}{2k_{t_p}} + \frac{N_p}{2k_d}\right)^{1/2} \tag{2}$$

where k_p is the propagation rate constant, k_i is the rate constant for the initiator decomposition, f is the initiator efficiency, V_p is the total polymer volume, k_{t_p} is the termination rate constant, N_A is Avogadro's number, k_d is the rate constant for the desorption of radicals and all other variables are as defined earlier. This model differs from that proposed by Litt in that the main termination reaction is assumed to occur within the polymer phase and is conversion dependent (gel effect). Friis further proposed that the rate constant for desorption, k_d, is of the form

$$k_d = \frac{2D_p k_{tr_m} [M_p]}{N_A(2D_p + k_p [M_p] r^2)} \tag{3}$$

by assuming that only monomeric radicals are capable of escaping the particles. In eqn. (3) D_p is the diffusion constant for the monomeric radicals and k_{tr_m} is the rate constant for transfer to monomer. Since the diffusion constant should be dependent on the viscosity, which will change with conversion, the diffusion constant in eqn. (3) was assumed to behave according to the relation

$$D_p = D_p^0 \exp \frac{-\beta x \rho_m (1-\alpha)}{(1-x)\rho_p + \alpha x \rho_m} \tag{4}$$

where x is the fractional conversion, ρ_m and ρ_p are the monomer and polymer densities, respectively, D_p^0 is the diffusion coefficient at zero conversion and α and β are parameters related to the free volume fractions in the particle.

In arriving at the values to be used for the two parameters α and β in his model, Friis used a curve fitting technique and the data from one of his experiments. This same technique was also used by Litt to determine the proper ionic strength dependence for his parameter C_2. Both authors did, however, present some data to indicate that the values they used were reasonable. In the latest paper concerning the aqueous phase termination model,[6] Stannett has examined both models and determined that the higher ionic strengths used in the work by Friis should cause a polymer phase termination. It is also suggested that the radical desorption rate constant in the model proposed by Litt should be conversion dependent. However, to do this would destroy one of the arguments advanced in support of the aqueous phase termination model, namely that the model fits the reaction rate data with no adjustment parameters. To include this factor would involve a curve fitting technique, such as that used by Friis, to model the conversion dependency of the desorption rate constant. The other factor which Stannett claims may have caused the differences between the results observed by Litt and those observed by Friis, the intensity of agitation, was shown earlier not to have an effect on the polymerization reaction.

All of the changes proposed by Stannett, however, as well as the original models do not change the one factor which remains unaccounted for in all the models, namely the changing behavior of the particle number with respect to the initiator concentration and the effect this has on the overall reaction order. Equations (1) and (2) both indicate that either model has only an explicit half-order dependence on the initiator concentrations. For the polymer phase termination model proposed by Friis, eqn. (2), the second term in the parentheses is usually much larger than the first. Thus if the particle number were to be directly proportional to the initiator concentration, the overall reaction would be first order with respect to the initiator. If the particle number was independent of the initiator concentration the overall reaction would exhibit a half-order dependence as was observed by Friis. The same holds true for the aqueous phase termination model, eqn. (1). In the paper by Stannett *et al.*[6] it is claimed that the limiting form of eqn. (1) at high ionic strengths also exhibits a half-order dependence on the initiator concentration. This limiting process also predicts that the particle number should be proportional to the particle diameter to the -3 power and the particle diameter should be proportional to the $0{\cdot}13$ power of the initiator concentration. Thus the particle number should be proportional to the $-0{\cdot}39$ power of the initiator. This prediction does not agree with the experimental observations reported in the earlier paper,[5] where the particle number was experimentally observed to depend on the initiator concentration to the 1·0–1·2 power. Thus none of the modifications to the original aqueous phase termination model proposed by Stannett *et al.*, nor the original model itself, can correctly account for the effects of the initiator on the particle number. The model proposed by Friis says nothing of this dependence.

Particle number model

Because of the relatively high solubility of the monomer in water, vinyl acetate emulsion polymerizations are assumed to start as an aqueous phase solution polymerization. As the polymer chains build in the aqueous phase they ultimately reach some critical length, where the chains coil on themselves and precipitate from the aqueous phase. This gives rise to the generation of new particles and the start of the emulsion polymerization process. This homogeneous generation of particles has been postulated for many other monomers[9] and usually results in an initial particle number greatly in excess of that observed during the bulk of the polymerization. These particles are not stable, however, as the extremely large surface area of the particles completely depletes the emulsifier. The particles then undergo a limited coagulation, reducing the surface area and freeing emulsifier until a stable particle population is obtained. The fact that emulsifier micelles do not enter into this homogeneous mechanism of particle generation is probably the reason that the emulsion polymerization of vinyl acetate does not exhibit a strong dependence on the emulsifier concentration.

The particle number behavior observed in Figs. 3, 4 and 5 can now be explained in terms of this homogeneous particle generation mechanism and the DLVO theory of colloidal stability.[10] At low initiator concentrations, which corresponds to low ionic strengths in the standard system, any particles generated are more likely to survive as the low ionic strength increases the stability of the particles. As the initiator concentration increases, the rate of particle initiation also increases which, for stable particles, should lead to an increase in the particle number. As the initiator level increases still further, the particle stability decreases, until a point is reached where the tendency of the particle number to increase with increasing initiator concentration is offset by the decreasing particle stability caused by higher ionic strengths. The results shown in Fig. 5 would indicate that for vinyl acetate this seems to occur at about $1{\cdot}0 \times 10^{-3}$ M initiator. The further increase in particle number with initiator concentration at $4{\cdot}0 \times 10^{-3}$ M initiator at 50°C, shown in Fig. 3, may be a result of a surface condensation of the emulsifier on the particle surface as hypothesized by Stannett *et al.*[6]

The true behavior of the particle number with conversion is very complex. Although not shown in Figs. 3 and 4 the particle number has been observed to peak at about 10^{21} particles/liter H_2O during the initial phases of the reaction before dropping down to the 10^{17} range and slowly increasing with conversion from that point. In order to determine if this particle number behavior could be responsible for the variability in previous investigations, a simple linear model was developed for use in simulating the conversion versus time histories for polymerization reactions. To do this the initial peak in particle number may be ignored and a model of the form

$$N_p = a + bx \tag{5}$$

developed where the two parameters a and b are functions of initator level, ionic strength, temperature, emulsifier level and type, and x is the fractional conversion. While the peak in particle number is real, it lasts only a short time, so that ignoring its existence should not lead to large errors in predicting conversion. To determine the behavior of the parameters a and b a set of designed experiments, as shown above the dashed line in Table II, was performed and the results analyzed using a multiple linear regression program, to obtain a model for predicting the particle number behavior. Since such models are in general only applicable within the range of the variables used in developing the model, the original eight experiments in the designed set were supplemented with an additional five experiments to extend the results

TABLE II
Designed set of experiments for vinyl acetate emulsion polymerization

Experiment	[I] $M\times10^3$	[S] $M\times10^2$	Ionic strength $M\times10^3$	T °C	Reaction rate g mole/liter H_2O s $\times10^4$	Particle number No./liter $H_2O\times10^{-17}$ Intercept	Slope
1	4·0	1·7	30·0	50	12·35	1·221	0·8216
2	0·25	6·9	30·0	50	2·13	0·7403	1·160
3	0·25	1·7	3·0	50	2·10	1·108	1·105
4	0·5	6·9	30·0	60	9·77	1·128	1·301
5	0·125	1·7	30·0	60	4·05	0·8520	0·8147
6	0·5	1·7	3·0	60	9·02	1·323	0·9463
7	0·125	6·9	3·0	60	3·53	0·5786	1·531
8	4·0	6·9	12·0	50	12·75	2·309	1·446
9	0·5	6·9	1·5	50	5·10	1·668	1·246
10	0·5	6·9	1·5	60	10·51	1·943	1·580
11	0·125	6·9	0·375	60	3·22	0·7096	1·343
12	1·0	6·9	3·0	60	14·92	2·106	1·559
13	1·0	6·9	3·0	50	6·16	1·558	0·8370

to lower ionic strengths. The result of a multiple linear regression on the results of all 13 experiments led to the following equation describing the behavior of the particle number during a vinyl acetate emulsion polymerization:

$$N_p\times10^{-17}=18{\cdot}64+4{\cdot}972\times10^8 fk_i\,[I]-1{\cdot}081\times10^3\,(\text{ionic strength})$$
$$-5{\cdot}405\times10^{-2}\,T-1{\cdot}1\times10^{10} fk_i\,[I]\,(\text{ionic strength})$$
$$+3{\cdot}276\,(\text{ionic strength})\,T+x(0{\cdot}7847+14{\cdot}63\,[S])$$

where N_p is the weight average particle number (number/liter H_2O), T is temperature (K) and all other variables are as defined earlier. The same set of experiments was repeated using sodium cetyl sulfate as the emulsifier, and no statistically significant differences between the results could be observed. Thus, the model should apply for all vinyl acetate emulsion polymerizations stabilized by sodium alkyl sulfates.

Comparison of reaction models with experimental data

Equation (6) may now be used with either the aqueous phase termination model (eqn. (1)) or the polymer phase termination model (eqn. (2)) to predict the conversion versus time history for a vinyl acetate emulsion polymerization. The data from one experimental run at 50°C and one at 60°C were used to determine the values of the parameters α and β, and C_2 to be used in the respective models. From this curve fitting the values of α and β determined in this investigation were 0·42 and 3·69 while those reported by Friis were 0·3 and 3·2. The value for C_2 was determined to be $2{\cdot}609 \times 10^{-12}$ mole s cm^2/liter while Litt reports a value of $4{\cdot}25 \times 10^{-12}$ to $1{\cdot}3 \times 10^{-12}$. Thus the values determined in this investigation are in agreement with the previous results. Using the particle number model developed above, the values of the parameters reported above and literature values for the other rate constants, both the aqueous phase termination model and the polymer phase termination model were numerically integrated to yield conversion versus time histories at both 50°C and 60°C. The results of such an integration at 50°C are shown in Fig. 6. The solid line represents the predictions from the polymer phase termination model (eqn. (2)) while the intermittent dashed line represents the results from the aqueous phase termination model (eqn. (1)). As can be seen in Fig. 6 the polymer phase model does a much better job of predicting the results than does the aqueous phase termination model. The dashed curve shown at high conversions represents the results that would be predicted by the polymer phase termination model if the gel effect were

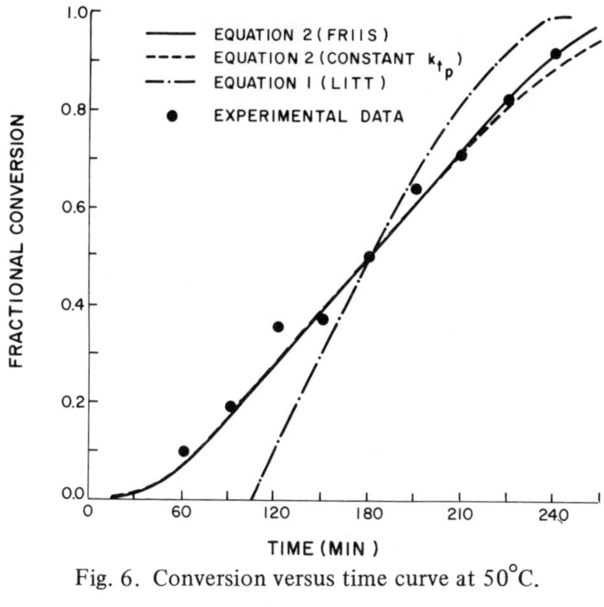

Fig. 6. Conversion versus time curve at 50°C.

ignored. The fact that the retardation of the termination reaction has such a minor effect indicates that the reaction rate is controlled mainly by the radical desorption rate for most of the conversion versus time history. The results of a similar integration at 60°C are shown in Fig. 7. Both models now are in close agreement, although the polymer phase termination model shows a slightly better fit at both the high and low conversion end. Again the gel effect is not important.

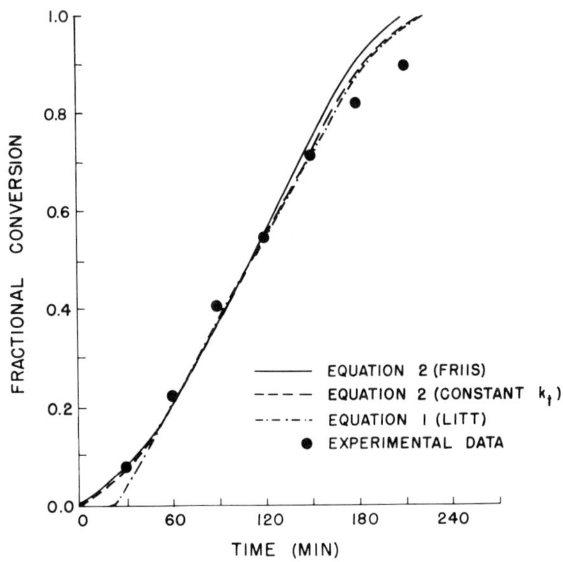

Fig. 7. Conversion versus time curve at 60°C.

It is interesting to note that the particle number model, presented as eqn. (6), predicts that the particle number should decrease as the ionic strength increases. Yet the reaction rate increases with increasing ionic strength, rather than decreasing as would be expected from the lower particle number. This increase in the reaction rate is due to the effect of the ionic strength on the partitioning of the monomer. A recent investigation has shown that the concentration of monomer in the polymer particles should increase with increasing ionic strength.[11] Using this data to modify the monomer concentration in the polymer phase, the conversion versus time history for one of the polymerizations with added salts was predicted. While the polymerization rate was observed to increase, the increase was only approximately half of the increase observed experimentally. The difference between the predicted behavior and the observed behavior may have arisen from the fact that the ionic strength may also cause a decrease in the escape of the monomeric free radicals from the particles. This would tend to increase the free radical concentration within the particles and increase the polymerization rate.

CONCLUSIONS

The main shortcoming of all recent reaction models for the emulsion polymerization of vinyl acetate has been their failure to account for the variable nature of the reaction rate and particle number with respect to the initiator concentration. This investigation indicates that this variability may be due to changes in the ionic strength of the aqueous phase, caused by the initiator, affecting the particle number during the polymerization. To account for these changes an empirical model has been developed to predict particle number behavior. When this model is used in conjunction with a polymer phase termination model developed by Friis it accurately predicts the conversion versus time history at all temperature, initiator and ionic strength levels. Further work is still required, however, to accurately determine the effect of ionic strength on the escape of monomeric free radicals from the polymer particles and to develop a theoretically sound particle number model.

REFERENCES

1. R. L. Zollars, *J. Appl. Polym. Sci.,* **24**, 1353 (1979).
2. N. Friis and L. Nyhagen, *J. Appl. Polym. Sci.,* **17**, 2311 (1973).
3. W. J. Priest, *J. Phys. Chem.,* **56**, 1077 (1952).
4. R. Patsiga, M. H. Litt and V. T. Stannett, *J. Phys. Chem.,* **64** 801 (1960).
5. M. H. Litt, R. Patsiga and V. T. Stannett, *J. Polym. Sci., A-1,* **8**, 3607 (1970).
6. V. T. Stannett, A. Klein and M. H. Litt, *Br. Polym. J.,* **7**, 139 (1975).
7. D. H. Napper and A. E. Alexander, *J. Polym. Sci.,* **61**, 127 (1962).
8. J. Ugelstad, P. C. Mork, P. Dahl and P. Rognes, *J. Polym. Sci., C.,* **27**, 49 (1959).
9. R. M. Fitch and C. H. Tsai, *Polymer Colloids,* Plenum, New York, 1971, p. 73ff.
10. A. S. Dunn and C. H. Chong, *Br. Polym. J.,* **2**, 49 (1970).

11. A. Klein, C. H. Kuist and V. T. Stannett, *J. Polym. Sci. Chem.*, **11**, 2111 (1973).
12. A. S. Dunn and P. A. Taylor, *Makromol. Chem.*, **83**, 207 (1965).
13. M. Nomura, M. Harada, K. Nakagawara, W. Eguchi and S. Nagata, *J. Chem. Eng. Jpn.*, **4**, 160 (1971).
14. D. Gershberg, *AIChE Int. Chem. Symp. Ser.*, **3**, 4 (1965).

4

Synthesis Kinetics and Characterization of Poly (Vinyl Acetate) – Molecular Weight and Long Chain Branching Development

A. E. HAMIELEC
Department of Chemical Engineering, McMaster University, Hamilton, Canada

SUMMARY

This chapter contains a review of recent developments concerning the kinetics of molecular weight development of poly(vinyl acetate) in solution, bulk and emulsion polymerization. This includes a discussion of analytical techniques based on liquid size exclusion chromatography for the measurement of frequency of long chain branching of poly(vinyl acetates) as a function of molecular weight. Kinetic models which account for transfer to monomer and polymer and consider terminal double bond reactions have an extensive range of validity in the emulsion polymerization of vinyl acetate.

The more recent investigations of the kinetics of synthesis and characterization of poly(vinyl acetate) (PVAc) have been reported by Graessley and co-workers,[1–8] and others,[9–14] and by Hamielec and co-workers.[15–20] The work of Graessley and co-workers involved solution and bulk polymerization of vinyl acetate at very low rates of initiation with a measurement of number average molecular weight by osmometry and weight average molecular weight by classical light scattering at different levels of conversion. Saponification and reacetylation were also done to find the location of long branches (α and β carbons on main chain and acetate group). The principal aim of these investigations was to establish the validity of a proposed kinetic mechanism which gives the development of $\bar{M}_N$, $\bar{M}_W$ and $\bar{B}_N$ (the number of long chain branches per polymer molecule) as a function of conversion. It was found that the proposed kinetic mechanism was consistent with measured $\bar{M}_N$ and $\bar{M}_W$ up to intermediate conversion levels for very low initiation rates and these permitted the numerical estimation of the kinetic rate constant ratios C_M, C_P and K. These ratios involve the propagation constant in the denominator and the transfer to monomer, transfer to polymer and terminal double bond reaction rate constants in the numerator, respectively. The kinetic equations with these known parameters were then solved to predict $\bar{B}_N$ as a function of conversion. No attempt was made to measure $\bar{B}_N$. The saponification and reacetylation studies revealed that about 70% of the long branches were on the acetate group.[5] Because the number of long branches per polymer molecule is very small ($\bar{B}_N \sim 1$–2) with negligible consumption of sites for transfer to polymer and terminal double bond reaction, one can argue that the ratio of the number of long branches on the main chain to that on the acetate group would not change with conversion. The activation energies of these branching reactions differ and one would therefore expect the ratio to vary with temperature. Hamielec and co-workers investigated the synthesis kinetics of PVAc using both bulk and emulsion polymerization. The kinetic equations for molecular weight and branching development after Graessley were modified somewhat for the case of emulsion polymerization.[15,16] It is of interest to note that commercial polymerization rates can be obtained in emulsion polymerization at much lower rates of initiation than in solution or bulk polymerization. The implication is that the kinetic scheme after Graessley is of much more practical value for emulsion polymerization and that it should be possible to accurately predict molecular weight and branching development in commercial emulsion polymerizations. This has been borne out by the investigations of the emulsion polymerization of vinyl acetate using sodium lauryl sulphate as emulsifier.[15,16] The kinetic model would not apply however when polyvinyl alcohol suspending agents are used, as chain transfer to suspending agent leads to insoluble copolymer. Hamielec and co-workers used gel permeation chromatography (GPC) with a low angle laser light scattering detector (LALLS)[17,18] and the molecular weight and branching distribution method (MWBD)[20] to investigate molecular weight and branching development. In a very recent and unpublished investigation of PVAc mechanical degradation and characterization by Jenkins and Porter[21] the following was reported. PVAc of $\bar{M}_W = 750\,000$, $\bar{M}_W/\bar{M}_N = 3{\cdot}85$ and $\bar{B}_N = 2{\cdot}2$ was mechanically (high speed stirring) and chemically (saponification and reacetylation) degraded to investigate the effect of shear on the branching distribution. Long chain branching was measured by GPC with a LALLS detector. It was concluded that (i) the branches through the acetate group are long and are ruptured preferentially on shearing, (ii) the branches through the α- and β-carbon are not broken on shearing, (iii) the extent of long branching through the acetate group is about 67% of total branching, (iv) on shearing, 80% of the decrease in molecular weight results from the main chain scission, and (v) the PVA derived from branched PVAc contains a smaller but nevertheless significant amount of branching.

Other workers who have reported the use of GPC and a LALLS detector to characterize polymers with long chain branching include Hamielec and co-workers[17–20] and Axelson and Knapp.[22]

KINETICS OF MOLECULAR WEIGHT AND BRANCHING DEVELOPMENT

Bulk and solution polymerization

If one assumes that the amount of polymer produced by termination reactions is negligible (very low rates of initiation) and that molecular weight and branching development is controlled by transfer to monomer, transfer to polymer and terminal double bond reactions, one can write the following set of equations, after Graessley, to describe molecular weight and branching development in bulk polymerization.

$$\frac{dQ_0}{dX} = C_M - \frac{KQ_0}{1-X} \tag{1}$$

$$\frac{dQ_1}{dX} = 1 \tag{2}$$

$$\frac{dQ_2}{dX} = 2\left(1 + \frac{KX}{1-X}\right)\frac{1 + \dfrac{C_P Q_2}{1-X} + \dfrac{KX}{1-X}}{C_M + C_P\dfrac{X}{1-X}} \tag{3}$$

$$\frac{d(Q_0\bar{B}_N)}{dX} = \frac{C_P X + KQ_0}{1-X} \tag{4}$$

$$\bar{\lambda}_N = 1000 M_R \bar{B}_N / \bar{M}_N \tag{5}$$

$$\bar{M}_N = \frac{M_R Q_1}{Q_0} \tag{6}$$

$$\bar{M}_W = \frac{M_R Q_2}{Q_1} \tag{7}$$

$$C_M = \frac{k_{fm}}{k_p}; \qquad C_P = \frac{k_{fp}}{k_p}; \qquad K = \frac{k_p^*}{k_p} \tag{8a, b, c}$$

where X is the conversion; Q_0, Q_1 and Q_2 are the zeroth, first and second moments of the frequency distribution of molecular weights; $\bar{B}_N$ is the number average number of long chain branches per polymer molecule; M_R is the molecular weight of the monomer repeat unit; $\bar{\lambda}_N$ is the number of long branches per 1000 monomer units; k_{fm} and k_{fp} are transfer constants to monomer and polymer, respectively; k_p is the propagation constant; and k_p^* is the terminal double bond reaction constant.

The usual initial conditions are, at $X = 0$,

$$Q_0 = Q_1 = Q_2 = \bar{B}_N = 0$$

Given values for C_M, C_P and K, these equations can be readily solved using standard numerical techniques. Polymer produced at low conversions should be linear and have the most probable distribution with a polydispersity of two. If $K = 0$, $\bar{M}_N$ would be independent of conversion, while $\bar{M}_W$ and $\bar{B}_N$ would increase with conversion due to transfer to polymer with concomitant long chain branching. All polymer molecules have a terminal double bond and if $K \neq 0$, $\bar{M}_N$ will increase with conversion, but much less so than $\bar{M}_W$, of course. Terminal double bond reactions produce long branches on the main chain. It is of interest to consider whether chain entanglements might affect the reactions which control molecular weight and branching development. On the basis of size of reacting molecules and intrinsic reactivity, terminal double bond reactions might be expected to become diffusion controlled. The terminal double bond reaction is a pseudo propagation with a rate constant $k_p^* \sim 10^3$ liter/g mole s and the end of the macroradical must find the terminal double bond at the end of a polymer chain, a rather difficult task in an entangled polymer matrix. There is experimental evidence that terminal double bond reactions do become diffusion controlled at higher conversions.[15] A solution of eqns. (1)–(7) for a typical set of kinetic parameters is shown in Fig. 1. There is a modest increase in $\bar{M}_N$ due to terminal double bond reactions. At high conversions $\bar{B}_N$ may reach 2–3 long branches per

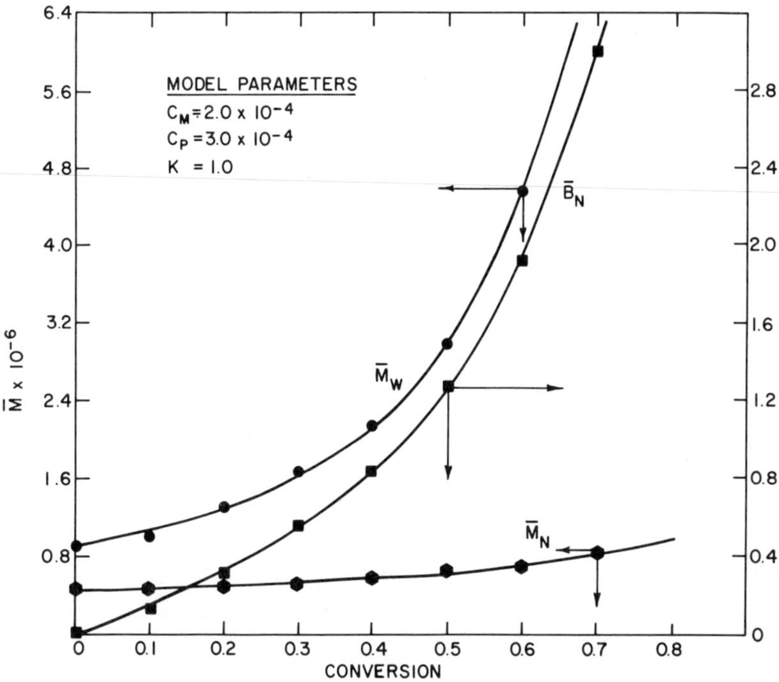

Fig. 1. Solution of kinetic equations for LCB in bulk VAc polymerization.

polymer molecule and of course the large increase in $\bar{M}_W$ is due entirely to long chain branching. It can be shown by neglecting the effect of terminal double bond reactions on Q_0 in eqn. (1) that transfer to polymer and terminal double bond reactions produces about the same number of long branches. A solution of eqn. (1) shows that $Q_0 = C_M X$ and if one now substitutes for Q_0 in the right-hand side of eqn. (4) one gets eqn. (4a)

$$\frac{d(Q_0 \bar{B}_N)}{dX} = (C_P + KC_M)\left(\frac{X}{1-X}\right) \tag{4a}$$

In other words the contribution to the total number of long branches is 60% due to transfer to polymer and 40% due to terminal double bond reactions. Graessley *et al.*[5] and Jenkins and Porter[21] have shown that about 70% of the long branches are on the acetate group. This means that at least 70% of the long branches must be formed by transfer to polymer as some transfer to α and β hydrogens on the main chain might occur. One therefore might expect a smaller value of K than unity as well as the possibility that K should

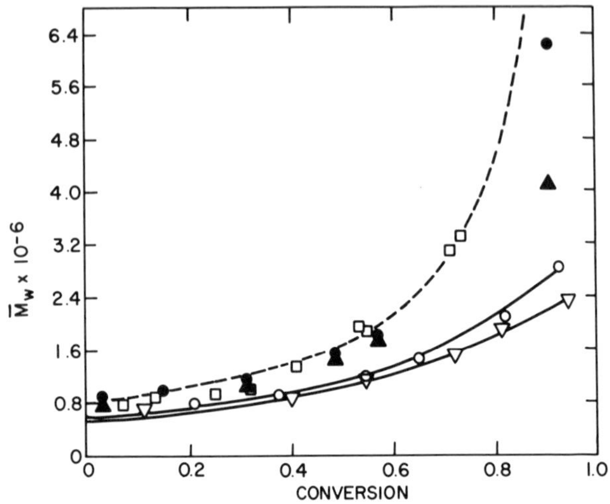

Fig. 2. A plot of the $\bar{M}_W$ versus percent conversion for PVAc synthesized at 60°C and three levels of initiator concentrations. ● – [AIBN] = 10^{-5} g mole/liter, LALLS offline; ▲ – [AIBN] = 10^{-5} g mole/liter, LALLS/GPC online data; ○ – [AIBN] = 2×10^{-3} g mole/liter, LALLS offline data; ▽ – [AIBN] = 4×10^{-3} g mole/liter, LALLS offline data; □ – Graessley's[5] data. – – – theory ($C_M = 2 \times 10^{-4}$, $C_P = 1{\cdot}5 \times 10^{-4}$, $K = 0{\cdot}56$).

decrease with conversion as polymer chains become highly entangled. Some $\bar{M}_W$ data showing the effect of initiator level on molecular weight development are shown in Fig. 2. It is clear that for solution and bulk polymerization, the kinetic model (eqns. (1)–(8)) can apply only at very low rates of initiation. The decrease of $\bar{M}_W$ with initiator level indicates that termination reactions are contributing significantly to polymer production. The analytical techniques used to measure $\bar{M}_W$ will be discussed later. It is of interest at this point to discuss the effect of inert solvent, solvent with some chain transfer activity and of a chain transfer agent on molecular weight and branching development. A recent study by Graessley and co-workers[23] employed *t*-butanol as a solvent with some chain transfer activity. An inert solvent will lower the polymer concentration at all conversion levels and thereby reduce the rate of transfer to polymer and terminal double reactions. $\bar{M}_N$ will therefore increase to an even smaller extent at high conversions. $\bar{B}_N$ and $\bar{\lambda}_N$ will both be smaller at all levels of conversion. A solvent with some chain transfer activity will reduce $\bar{M}_N$ and $\bar{M}_W$ at all conversion levels. $\bar{B}_N$ and $\bar{\lambda}_N$ will again both be smaller at all conversion levels with $\bar{B}_N$ falling more because of the additional decrease in $\bar{M}_N$ and increase in the number of polymer molecules.

It is interesting to note that Graessley and co-workers[23] observed a solvent effect on C_P. The propagation constant k_p may change with solvent level; however, this should also have influenced C_M and K as well as C_P. A chain transfer agent would increase the number of polymer molecules by producing those without a terminal double bond. This would reduce the production of long branches via terminal double bond reactions and thus the total number of long branches. $\bar{\lambda}_N$ would be smaller and $\bar{B}_N$ even more so because of the larger number of polymer molecules. A convenient strategy to maintain relatively high molecular weights with reduced long chain branch lengths at high conversions would be to add the chain transfer agent at high conversions. This would have the effect of producing shorter long chain branches on a long main chain produced earlier in the reaction.

Emulsion polymerization

In stages I and II of emulsion polymerization the monomer or polymer concentrations are constant and the appropriate equations for molecular weight development follow.[15,16]

$$\bar{M}_N = \frac{M_R}{C_M}\left(1 + \frac{KX_C}{1 - X_C}\right) \tag{9}$$

$$\bar{M}_W = \frac{2M_R\left(1 + \dfrac{KX_C}{1 - X_C}\right)^2}{\left[C_M + \dfrac{C_P X_C}{1 - X_C} - \left(\dfrac{2C_P X_C}{1 - X_C}\right)\left(1 + \dfrac{KX_C}{1 - X_C}\right)\right]} \tag{10}$$

$$\bar{B}_N = \left(\frac{X_C}{1 - X_C}\right)\left[\frac{C_P}{C_M}\left(1 + \frac{KX_C}{1 - X_C}\right) + K\right] \tag{11}$$

where X_C is the weight fraction of polymer in monomer or the conversion where stage II ends. To calculate molecular weight and branching development in stage III one must solve eqns. (1)-(8) using initial values for $\bar{M}_N, \bar{M}_W$ and $\bar{B}_N$ at $X = X_C$ given by eqns. (9)-(11). In emulsion polymerization the branching frequencies are higher at the same conversion levels because polymerization in the polymer particles always occurs at a polymer concentration considerably greater than zero. Kinetic parameters observed for emulsion polymerization follow.[15,16]

$$C_M = 0{\cdot}2\ \exp(-4290/RT) \tag{12}$$

$$C_P = 0{\cdot}076\ \exp(-3340/RT) \tag{13}$$

$$K = 0{\cdot}49 \tag{14}$$

At 60°C, $C_M = 3 \times 10^{-4}$, $C_P = 4{\cdot}8 \times 10^{-4}$, $K = 0{\cdot}49$ and $X_C = 0{\cdot}2$. In stages I and II, $\bar{B}_N = 0{\cdot}6$ already, an appreciable value at 20% conversion. It should also be mentioned that it has been shown that even for commercial initiation rates for the emulsion polymerization of vinyl acetate, a negligible amount of polymer is produced by termination reactions and that therefore as a result the above kinetic equations which describe molecular weight and branching development are valid. This is not so for solution and bulk polymerization.

CHARACTERIZATION OF BRANCHED POLY(VINYL ACETATE)

Gel permeation chromatography

When characterizing polymers with long chain branching such as poly(vinyl acetate), one should remember that the whole polymer is a mixture of linear polymer chains and polymer chains with one, two or more long branches per polymer molecule. The branch lengths on the average are the same as the average main chain length. If the polymerization temperature were allowed to increase with time and if a chain transfer agent were added at higher conversion levels, then the branches would be shorter than the main chains on the average. At any instant of time, therefore, the detector cell of a GPC would contain polymer molecules of differing size in solution and molecular weight. The different sizes are a direct result of peak broadening. If peak broadening were absent the polymer molecules would have the same size but would still differ in molecular weight. We are dealing here with complex polymers, polymers whose size in solution depends on both molecular weight and branching structure. Hamielec and Ouano[17] have generalized the universal molecular weight calibration curve after Benoit to include complex polymers. The ordinate is now $\ln[\eta]\bar{M}_N$ versus retention volume. An online viscometer detector response corrected for peak broadening along with the generalized universal molecular weight calibration curve provides a measure of $\bar{M}_N(v)$ across the chromatogram of the branched polymer sample. The $\bar{M}_N$ of the whole polymer may then be found with integration across the chromatogram according to

$$\bar{M}_N = 1 \Bigg/ \int_0^\infty \frac{F_N(v)\,\mathrm{d}v}{\bar{M}_N(v, uc)} \tag{15}$$

where $F_N(v)$ is the normalized mass detector response (DRI, UV or IR) and $\bar{M}_N(v, uc)$ is the number average molecular weight of the contents of the detector cell. $\bar{M}_N(v, uc)$ is found by correcting $\bar{M}_N(v)$ for peak broadening using equations for complex polymers.[19] In the absence of an online viscometer detector, one may employ the molecular weight and branching distribution method (MWBD) proposed by Foster *et al.*[20] Examples of two branched PVAc samples characterized by MWBD are shown in Figs. 3 and 4. The measurement of the whole polymer $\bar{M}_W$ may be done offline with a LALLS, or online with GPC coupled with a LALLS detector.[17,18,21] With the latter combination, the $\bar{M}_W$ of the whole polymer may be found with integration across the chromatogram according to

$$\bar{M}_W = \int_0^B F_N(v)\, \bar{M}_W(v, uc)\, \mathrm{d}v \tag{16}$$

where $\bar{M}_W(v, uc)$ is the weight average molecular weight of the contents of the detector cell and is given by the combination of LALLS and a mass detector.

Several workers [21,22] have recently attempted to use $\bar{M}_W(v, uc)$ measured by GPC coupled with LALLS and a mass detector to estimate long chain branching. There are certain difficulties with this procedure and they will now be briefly discussed. The actual procedure used follows. The universal molecular weight calibration curve based on polystyrene and $\bar{M}_W(v, uc)$ values are used to generate $[\eta]_b(v)$ values, intrinsic viscosities across the chromatogram of the branched polymer sample. The $[\eta]_l(v)$ are found using appropriate Mark-Houwink constants for linear polymer (say linear PVAc). The ratio of $\{[\eta]_b(v)\}/\{[\eta]_l(v)\}$ and a branching model[24] with a suitable ϵ may be used to estimate $\bar{B}_N(v)$, the number average number of branch points per polymer molecule for the contents of the detector cell at retention volume v. The whole polymer $\bar{B}_N$ and $\bar{\lambda}_N$ may then be calculated using

$$\bar{B}_N = \bar{M}_N \int_0^\infty \frac{\bar{B}_N(v)\, F_N(v)\, \mathrm{d}v}{\bar{M}_N(v)} \tag{17}$$

$$\bar{\lambda}_N = 1000 M_R \frac{\bar{B}_N}{\bar{M}_N} \tag{18}$$

Because $\bar{M}_N(v)$ and $\bar{M}_N$ are not known, it is assumed that $\bar{M}_N(v) = \bar{M}_W(v, uc)$. Many of the steps used above are thus incorrect and could lead to significant error in $\bar{B}_N$. The choice of ϵ in the branching model ($\epsilon = 0{\cdot}5$ for star-type

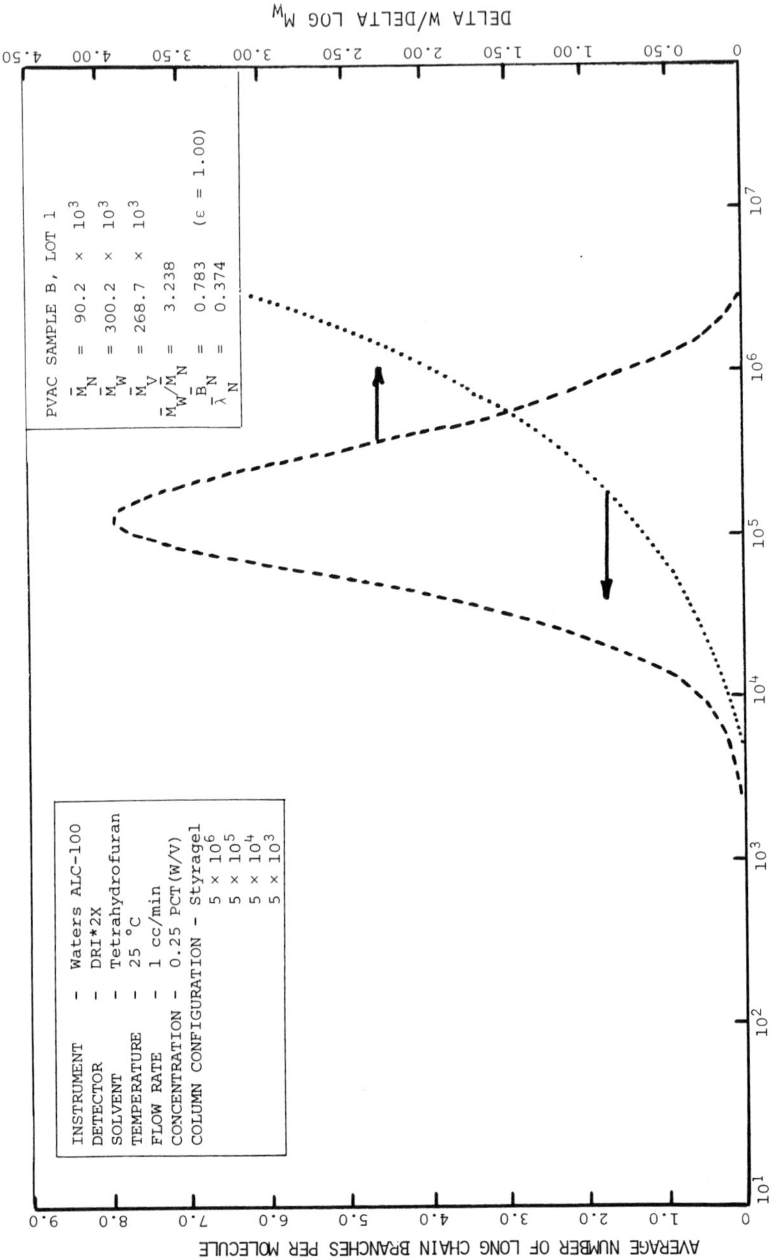

Fig. 3. Characterization of poly(vinyl acetate)—gel permeation chromatography and molecular weight and branching distribution method.[20] PVAc Sample B, Lot 1.

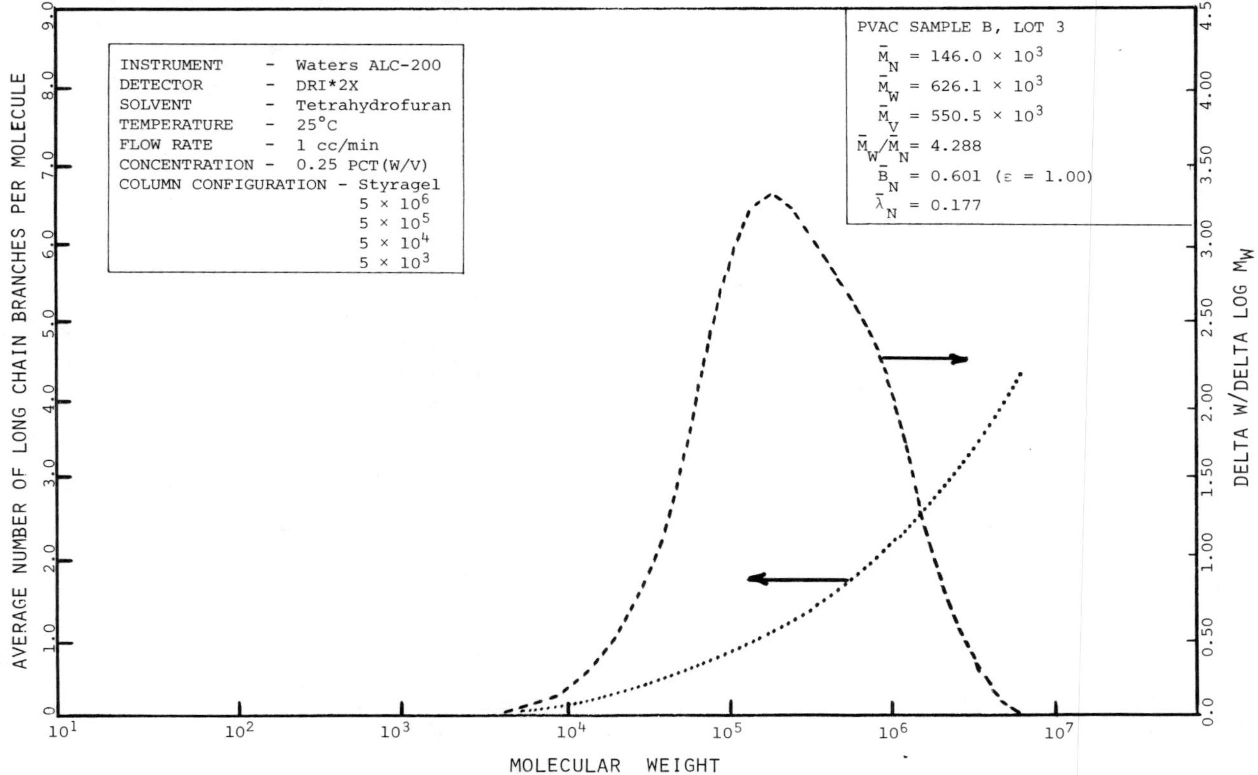

Fig. 4. Characterization of poly(vinyl acetate)—gel permeation chromatography and molecular weight and branching distribution method.[20] PVAc Sample B, Lot 3.

and $\epsilon = 1{\cdot}5$ for comb-type branching) can have a very large influence on the measured branching frequency. For example, changing ϵ from 0·5 to 1·2 changed $\bar{B}_N$ from 2·5 to 0·34.[20] ^{13}C NMR has been used to show that $\epsilon = 0{\cdot}75$ is an appropriate value for high pressure, low density polyethylene[20] while a consideration of synthesis kinetics and measurements of $\bar{M}_N$ and $\bar{M}_W$ lead to $\epsilon = 1{\cdot}0$ for poly(vinyl acetate). The setting of $\bar{M}_N(v) = \bar{M}_W(v, uc)$ would probably give $\bar{B}_N$ and $\bar{\lambda}_N$ which are too large. Peak broadening would give $\bar{M}_W(v, uc)$ which are smaller than $\bar{M}_W(v)$, the weight average molecular weight of the detector cell contents at retention volume, v. This is true at the leading edge of the chromatogram where most of the highly branched molecules reside and would tend to reduce the error associated with setting $\bar{M}_N(v) = \bar{M}_W(v, uc)$. There is an uncertainty in all the GPC approaches used to estimate long chain branching frequency. Branching models which relate the number of long branches per molecule to the ratio of intrinsic viscosities for branched and linear chains of the same molecular weight are being applied to the contents of the detector cell. The contents are not monodispersed in molecular weight. The polymer molecules may have the same size in solution but possibly very different molecular weights, particularly at the high molecular weight end of the chromatogram. Calibration of GPC via NMR, an absolute method of measurement of long chain branching frequency, should reduce the errors associated with the incorrect use of the Zimm–Stockmayer branching models.

REFERENCES

1. W. W. Graessley and H. M. Mittelhauser, *J. Polym. Sci., A2,* **5**, 431 (1967).
2. W. W. Graessley and J. S. Prentice, *J. Polym. Sci., A2,* **6**, 1887 (1968).
3. O. Saito, K. Nagasubramanian and W. W. Graessley, *J. Polym. Sci., A2,* **7**, 1937 (1969).
4. K. Nagasubramanian, O. Saito and W. W. Graessley, *J. Polym. Sci., A2,* **7**, 1955 (1969).
5. W. W. Graessley, R. D. Hartung and W. C. Uy, *J. Polym. Sci., A2,* **7**, 1919 (1969).
6. K. Nagasubramanian and W. W. Graessley, *Chem. Eng. Sci.,* **25**, 1559 (1970).
7. W. W. Graessley and E. S. Shinbach, *J. Polym. Sci. – Polym. Phys. Ed.,* **12**, 2047 (1974).
8. W. S. Park and W. W. Graessley, *J. Polym. Sci. – Polym. Phys. Ed.,* **15**, 85 (1977).
9. D. J. Stein, *Makromol. Chem.,* **76**, 170 (1964).
10. M. K. Linderman, *Vinyl Polymerization*, vol. 1, Part 1, G. E. Ham (ed.), Edward Arnold, London, 1967, Chapt. 4.

11. W. C. Uy, *J. Polym. Sci., A2,* **7**, 1919 (1969).
12. G. Odian, *Principles of Polymerization,* McGraw-Hill, New York, 1970, p. 219.
13. S. Nazakura, Y. Morishima and S. Murahashi, *J. Polym. Sci. – Polym. Chem. Ed.,* **10**, 2853 (1972).
14. P. A. Small, *Adv. Polym. Sci.,* **18,** 1 (1975).
15. N. Friis, D. Goosney, J. D. Wright and A. E. Hamielec, *J. Appl. Polym. Sci.,* **18,** 1247 (1974).
16. N. Friis and A. E. Hamielec, *J. Appl. Polym. Sci.,* **19,** 97 (1975).
17. A. E. Hamielec and A. C. Ouano, *J. Liquid Chromatography,* **1,** 111 (1978).
18. A. E. Hamielec, A. C. Ouano and L. L. Nebenzahl, *J. Liquid Chromatography,* **1,** 527 (1978).
19. A. E. Hamielec, *J. Liquid Chromatography,* **3,** 381 (1980).
20. G. N. Foster, A. E. Hamielec and T. B. MacRury, *Characterization of polymers with long chain branching – Development of the molecular weight and branching distribution method (MWBD),* in: *Size Exclusion Chromatography,* T. Provder (ed.), ACS Symp. Series no. 138, 1979, pp. 131–48.
21. R. F. Jenkins and R. S. Porter, *Characterization of branched poly(vinyl acetate) by gel permeation chromatography coupled with on-line low angle laser light scattering photometry*, unpublished manuscript, University of Massachusetts, Amherst, 1980.
22. D. E. Axelson and W. C. Knapp, *J. Appl. Polym. Sci.,* **25,** 119 (1980).
23. A. Chatterjee, K. Kabra and W. W. Graessley, *J. Appl. Polym. Sci.,* **21,** 1751 (1977).
24. B. H. Zimm and W. H. Stockmayer, *J. Chem. Phys.,* **18,** 1301 (1949).

5

Some Surface Chemical Effects on the Capture of Free Radicals – A Kinetic Parameter Analysis of the LPS Model

A. KLEIN
Emulsion Polymers Institute, Lehigh University, Pennsylvania, USA

and

V. T. STANNETT
Department of Chemical Engineering, North Carolina State University, USA

SUMMARY

The LPS kinetic model parameters were either experimentally determined or they were available in the literature. Using the kinetic model and statistical techniques, the only adjustable parameter, the oligomeric radical capture rate constant, was evaluated as a function of ionic strength and pH. The capture rate constant was decomposed into a stability factor term and an ionic strength independent rate constant. The stability factor, as a function of ionic strength, exhibited a deep minimum at 0·014 ± 0·002 M ionic strength. The rate dependence on initiator was found to have a 0·8 power dependence below 0·014 M ionic strength and 0·5 power dependence above 0·014 M ionic strength. This correlates well with data from recent literature, which reports at low ionic strength a 0·88 ± 0·14 power dependence and at high ionic strength a 0·58 power dependence.

The pH effect on the stability factor was suggested in terms of association between the acetic acid and hydroxyl at the particle surface, both of which are the products of polymer hydrolysis.

NOMENCLATURE

$\bar{a}$	the average organic to aqueous phase ratio
b	the fractional conversion at which the separate monomer phase disappears

k_0''	the least square smoothed value of k_0, against hydrogen ion concentration
k_1	the uncatalyzed persulfate decomposition rate constant
k_2	the hydrogen ion catalyzed persulfate decomposition rate constant
k_4	the rate constant for 'sweep-up of solubilized polymeric molecules' by the particles
k_4^o	the ionic strength independent component of k_4
k_5	the rate constant for propagation
k_6	the chain transfer rate constant
k_8	the rate constant for the butyrolactonyl radical reaction with monomer in the aqueous phase
$k_{10.2}$	the rate constant for termination in the aqueous phase between butyrolactonyl radical and 'solubilized polymeric radical'
k_i	the initiation rate constant
$[M]$	the molar monomer concentration in the system
$[M]_0$	the initial molar monomer concentration in the system
$[M_{aq}]$	the molar monomer concentration in the aqueous phase
$[M_{aq}]_b$	$[M_{aq}^o]$ = the molar saturation monomer concentration in the aqueous phase
$[M]_t$	the molar monomer concentration in the system as a function of time
$[M_{aq}^D]$	the molar monomer concentration contributed by the diffused monomer from the aqueous phase, after the separate monomer phase disappeared
$[M_p]$	the molar monomer concentration in the polymer phase
$[M_p^o]$	the saturation molar monomer concentration in the polymer phase
$[M_p]_0$	the initial molar monomer concentration in the polymer phase
$[M_p]_b$	the molar monomer concentration in the polymer phase at the point of disappearance of the separate monomer phase
$[M_s]$	the equivalent molar monomer concentration existing as polymer seed
$[N_p]$	the molar particle concentration
R_p	the rate of polymerization
R_{pb}	the rate of polymerization at the point of separate monomer phase disappearance
RSS	the residual sum of squares
t	the time
t_b	the time at which the separate monomer phase disappears
W	the stability factor

INTRODUCTION

Vinyl acetate emulsion polymerization has been well studied in the past 15 years,[1-6] and while the deviation from the classical Smith–Ewart scheme[7] is well-established, there is no single model which explains all the trends. The state of the art for persulfate initiated polymerizations, in the presence of sodium lauryl sulfate, was critically reviewed.[8] Colloidal factors, while thought to be important in charged particle/radical collisions,[9] monomer inhibition into particles,[10] and in the homogeneous nucleation process,[11] were considered explicitly only by the Litt, Patsiga and Stannett (LPS) model.[4,8]

Recently, the LPS and Friis models were tested with new experimental data and were shown to be valid at low and high ionic strengths, respectively.[6] The two models predict different reaction rate dependence on the initiator level, first versus half-order dependence, and the two models explain much of the varied experimental data in the literature. The LPS model has a single adjustable parameter, the rate constant for radical capture by existing particles, while the Friis model contains two such parameters in describing the changing particle composition on the diffusivity of the monomer.

The LPS model is based on the existence of oligometric radicals in the aqueous phase onto which surfactant molecules have adsorbed such that they became dispersed and stabilized, similar in physical behavior to polyelectrolytes, but which undergo coagulation with increasing ionic strength.[12] The LPS model further postulates that the only means for a radical to enter the main locus of polymerization is for a monomer–polymer particle to capture the solubilized aqueous polymeric radical. Since both particle and radical are similarly charged by adsorbed surfactant, the capture is opposed by the repulsive interactions of the electrical double layers, and hence the capture rate constant is ionic strength dependent.

The LPS model was chosen to investigate the response of the adjustable parameter, the capture rate constant, to surface chemical perturbations such as ionic strength and pH. Statistical techniques were used to isolate the best estimate for the capture rate constant, from experimental data, which was decomposed into a rate constant and a stability factor term.[8] Accounting for the ionic strength effect on all the model variables yields a rate dependence on initiator to the 0·8 power at low ionic strength, and half-power dependence at high ionic strength. The effect of pH is complex and a hypothesis is put forward to explain it.

THE LPS KINETIC MODEL

Litt *et al.*[4] proposed that initiation is taking place in the aqueous phase to a degree of polymerization of 50 to 300, for an initiated monomer radical. It is further postulated that this happens because ionic soap is adsorbed on the polymer molecule, transforming it into a solubilized polymeric chain. Some of these chains are active and will be swept up by monomer–polymer particles.

Within the particles the monomer concentration is high and the captured polymeric radical will further grow or chain transfer to a monomer. This chain transfer is high for vinyl acetate. The monomer radical inside the particle will cyclize to a preferred butyrolactonyl radical, which is preferentially partitioned in the aqueous phase, where it grows into a chain, stabilized by adsorbed soap and swept up by a particle.

The termination step is postulated to take place mainly in the aqueous phase between polymeric chains and butyrolactonyl radicals. Thus, the essential features of the mechanism are: the polymerization or propagation step which mainly takes place inside the particle and the termination step which primarily occurs in the aqueous phase. The unique feature of this mechanism is seen in the importance of the rate constant for sweep-up of the solubilized aqueous polymer radicals by the particles.

The combination of assumptions mentioned above with the elementary kinetic steps yields the general rate expression:

$$R_p = \frac{k_5}{k_6}[k_i k_6 [C_i]\,[N_p]\,[M_p]]^{1/2}\left[1 + \frac{k_{10.2} k_6 [M_p]}{k_4 k_8 d^2\,[M_{aq}]}\right]^{-1/2}$$

where the symbols are defined in the Nomenclature.

The rate constant for the 'solubilized polymer radical' sweep-up, k_4, can be further decomposed into the relationship:

$$k_4 = \frac{k_4^o}{W}$$

where k_4^o from classical coagulation theory is evaluated to be:

$$k_4^o = \frac{kT}{3\eta}$$

and the stability factor W is a complex function of ionic strength, surfactant adsorption, and particle size.

Non-linear least square estimation of parameters K_1 and K_2

The integrated rate equation, when separate monomer phase is present, can be represented in the form:

$$\left[\frac{[M]_0 - [M] + [M_s]^{2/3}}{[M]_0 b + [M_s]} + K_2\right]^{3/2} - \left\{\left[\frac{[M_s]}{[M]_0 b + [M_s]}\right]^{2/3} + K_2\right\}^{3/2} = \frac{K_1}{[M]_0 b + [M_s]} t$$

where

$$K_1 = \frac{k_5}{k_6} [k_i k_6 [C_i] [M_p^o] [N_p]]^{1/2}$$

and

$$K_2 = \frac{k_{10.2} k_6 [M_p^o]}{k_4 k_8 d_b^2 [M_{aq}]_b}$$

where subscript zero indicates initial conditions, and d_b designates the size of the particle at the point of disappearance of the separate monomer phase. The fractional conversion at which monomer phase disappears is represented by b.

After the separate monomer phase disappears, the integrated rate equation is:

$$(1 + K_2)^{3/2} - \left[1 + K_2\left(\frac{[M_p]}{[M_p^o]}\right)^{1/2}\right]^{3/2} = \frac{3}{4} \frac{K_1 K_2}{\bar{a}[M_p^o]} (t - t_b)$$

where $\bar{a}$ is the average ratio of the organic to aqueous volumes, and t_b is the time when the separate monomer phase disappears.

Since the parameters K_1 and K_2 are constants within a polymerization experiment, b can be evaluated from simple mass balance, and while $[M]_0$, $[M]$, $[M_s]$, and t are measured quantities, the constant parameters are accessible from the experimental data.

This can be achieved by a search routine for the detection of the minimum in the residual sum of squares, RSS, in the two-parameters space, K_1 and K_2. The RSS is defined as:

$$\mathrm{RSS} = \sum_t [([M]_0 - [M]_t)_m - ([M]_0 - [M]_t)_c]^2$$

where subscripts m and c refer to measured and calculated conversions at time t. This RSS value can be evaluated as a function of K_1 and K_2; where the

RSS is a minimum, the grid value corresponds to the best values of K_1 and K_2 which fit the experimental data.

EXPERIMENTAL

Materials

The vinyl acetate monomer and sodium lauryl sulfate surfactant were obtained from Eastman Kodak. The latter was used without any further purification. The vinyl acetate was triply distilled by a spinning band column. The ionic strength was controlled by the addition of potassium sulfate solution, prepared from Merck's reagent crystals and distilled water. Potassium persulfate initiator solutions were made up by using Matheson, Coleman, and Bell's reagent crystals and distilled water.

Buffers, potassium dihydrogen phosphate, and sodium phosphate were Fisher and Baker reagent grade chemicals that were made up in distilled water.

Procedure

A small-volume, 30 cm^3, dilatometer was used for the rate measurements. The dilatometer was designed for the use of seeded emulsion polymerization, where degassing by freeze–thaw cycles must be avoided.[13] The monomer solubility in the polymer particles was determined by the vapor pressure method described elsewhere.[10] The persulfate decomposition rate was studied *in situ* in the polymerizing systems by polarographic means.[10,14]

Particle size distribution was measured with an electron microscope using a freeze-drying technique with carbon replication in order to harden the otherwise soft poly(vinyl acetate) particles in the electron beam.[15]

Results

Seed latex was prepared by γ-radiation of 151·4 g of vinyl acetate, and 10·9 g of sodium lauryl sulfate in 300 g of water. The pH right after preparation was 5·5 which gradually declined in a few days to 3·20 where it stabilized. The particle size distribution is shown in Fig. 1.

The reaction rate was found to be sensitive to the purity of the vinyl acetate obtained by distillation. The impurity was found to be acetone. The optimum purity obtainable with the spinning band column was 50 ppm of acetone. The effect of acetone on the rate is shown in Table I and a log–log plot of rate versus acetone content is shown in Fig. 2.

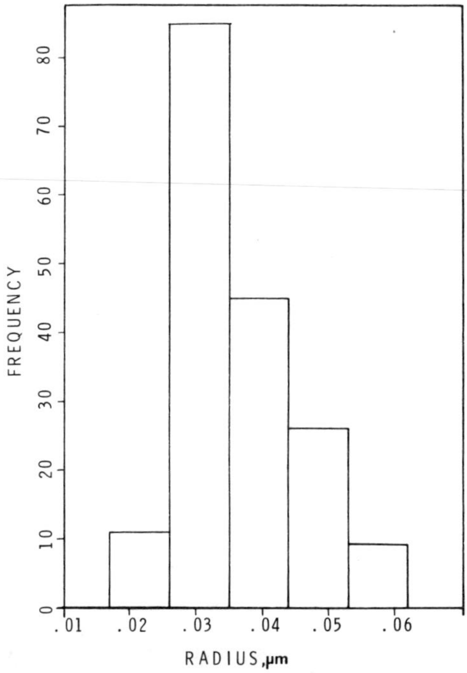

Fig. 1. Particle size histogram for seed latex.

TABLE I
Data on rate dependence on acetone at 60·2°C

Acetone ppm	*Weight[a] of water*	*Weight[a] of vinyl acetate*	*Weight of NaLS[b]*	*Rate[c]*
50	31·9590	4·7631	0·1309	3·416
208	32·3114	4·6308	0·1300	3·719
764	31·9923	4·6881	0·1300	3·663
3000	32·1016	4·6352	0·1303	3·293
7180	31·7908	4·7051	0·1292	3·128

[a] Units in grams
[b] Sodium dodecyl sulfate
[c] Initiator concentration = $5{\cdot}5 \times 10^{-4}$ M, in mole/liter min $\times 10^2$

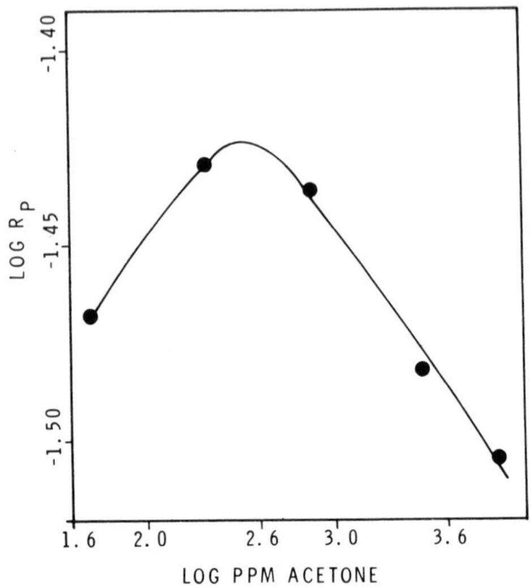

Fig. 2. Rate dependence on acetone content.

The rate dependence on ionic strength and pH was to be determined according to the grid of these two parameters as shown in Table II.

During the initial phase of the execution of the above plan, it became apparent that:

(i) at low ionic strength the buffer capacity was inadequate to keep the pH reasonably constant, while
(ii) at high ionic strength, even with the post-addition of 0·2% to 0·6% sodium dodecyl sulfate, the latex was unstable under polymerization conditions.

TABLE II
Experimental design grid in pH and ionic strength

pH	*Ionic strength*		
7·4	0·15	0·075	0·015
4·7	0·15	0·075	0·015
3·2	0·15	0·075	0·015

The alternate program, therefore, was clearly outlined by the limitations of the system. Experiments must take place at a lower than 0·075 M ionic strength. The experimental results, at 60·2°C, are shown in Table III.

TABLE III
Experimental results of polymerization rates of seed at varying ionic strengths and pH

Ionic strength	*Weight[a] of water*	*pH*		*Buffer[b]*	*Electrolyte[c]*	*Rate[d]*
		Initial	*Final*			
0·002	24·7169	3·70	3·40	–	–	3·839
0·008	24·7406	4·80	4·50	–	1	3·815
0·015	24·6890	4·45	4·35	2·5 ml	–	4·075
0·015	27·7088	5·50	4·95	–	2	4·014
0·045	24·7808	7·75	5·00	2·5 ml	3	3·222
0·045	24·6263	4·65	4·40	2·5 ml	4	3·218
0·075	24·6890	5·10	4·45	2·5 ml	5	2·981

[a] Units in grams
[b] Potassium dihydrogen phosphate solution
[c] 1 – 1·5 ml of 0·1 N sodium hydroxide
2 – 0·0178 g ammonium chloride + 2·5 ml of 0·1 N sodium hydroxide
3 – 3 ml of 0·1 N sodium hydroxide potassium sulfate solution
4 – 0·67 ml potassium sulfate solution
5 – 1·35 ml potassium sulfate solution
[d] Mole/min/1000 g $\times 10^2$

Note: Initiator – $7{\cdot}0 \times 10^{-4}$ M $K_2S_2O_8$
Weight of seed latex 2·54 g
Weight of monomer 4·6 g

The pH effects were studied at two levels of ionic strength. The pH effect is negligible while the ionic strength effect is quite significant. While the magnitude of change in the rate versus ionic strength is small, as expected from Patsiga's work,[16] the trend is well-defined and unexpected.

The monomer solubility in the polymer as a function of ionic strength was determined by the vapor pressure technique[10] and the data are given in Table IV.

The persulfate content of seed polymerized latexes was determined *in situ*, at three temperatures, by polarographic means at the end of the reaction. The experimental data are shown in Table V. The rate constants

TABLE IV
Derived solubility parameters as a function of ionic strength at 60·2°C

Ionic strength	*Polymer/solvent interaction parameter*	*Interfacial tension (dynes/cm)*	*Monomer solubility in polymer molar*
0·002	0·25	17·5	7·51
0·008	0·25	12·0	7·93
0·015	0·25	10·0	8·21
0·045	0·25	7·0	8·42
0·045	0·25	7·0	8·38
0·075	0·25	6·1	8·55

TABLE V
Experimental data on persulfate decomposition rate

Temperature in °C	$(H^+) \times 10^5$ M	*Ionic strength*	$k_0 \times 10^4\ min^{-1}$	$k''_0 \times 10^4\ min^{-1}$
65·0	31·60	0·002	49·83	49·83
65·0	0·16	0·008	12·20	12·63
65·0	0·50	0·015	13·34	12·90
60·2	28·20	0·002	36·83	36·86
60·2	0·04	0·045	7·32	7·90
60·2	3·98	0·045	10·25	9·63
49·8	20·20	0·002	14·79	14·83
49·8	2·00	0·008	2·68	3·00
49·8	5·02	0·015	4·52	4·16

H^+ – Hydrogen ion concentraton
k_0 – Decomposition rate constant, experimental
k''_0 – Decomposition rate constant, least square smoothed values against hydrogen ion concentration

were strongly dependent on the hydrogen ion concentration, which was calculated from the arithmetic average of the initial and final pH of the latex. This functional relationship was found to be linear for all three temperatures. The least square intercept, k_1, is the hydrogen ion concentration independent rate constant, and the slope, k_2, is the hydrogen ion dependent rate constant shown in Table VI, as a function of temperature.

TABLE VI
Temperature dependence of the decomposition rate constants during polymerization

Temperature in °C	$k_1 \times 10^4\, min^{-1}$	$k_2 \times 10^4\, min^{-1}$
65·0	12·38	1·19
60·2	6·69	1·07
49·8	1·26	0·67

The activation energies obtained by the least square method for the hydrogen ion uncatalyzed and catalyzed decompositions were found to be 33·2 kcal/mole and 8·5 kcal/mole, respectively.

Particle size data in terms of the derived moments of the distribution are shown in Table VII.

TABLE VII
Particle size data at 60·2°C

Ionic strength	$\bar{r}_n$	$\bar{r}_s$	$\bar{r}_w$	$\bar{r}_w/\bar{r}_n$	$\bar{V}_n \times 10^{16}$
0·002	0·0599	0·0720	0·0773	1·29	2·86
0·008	0·0702	0·0821	0·0869	1·24	4·40
0·019	0·0582	0·0648	0·0649	1·17	2·32
0·015	0·0639	0·0692	0·0712	11·11	2·95
0·045	0·0612	0·0661	0·0681	1·11	2·69
0·049	0·0604	0·0687	0·0724	1·20	2·58
0·075	0·0606	0·0668	0·0696	1·15	2·58

$\bar{r}_n$ – Number average radius
$\bar{r}_s$ – Surface average radius
$\bar{r}_w$ – Weight average radius
$\bar{V}n$ – Number average volume

Discussion

In order to evaluate the adjustable parameter, the rate constant for oligomeric radical capture by particles, first the kinetic parameters K_1 and K_2 were evaluated. The experimental conversion-time data and kinetic model were used to construct the residual sum of squares, RSS. Instead of using some hill climbing techniques to minimize RSS in K_1 and K_2, it was decided that the surface within a reasonable experimental space would be investigated. A computer program to calculate the RSS values for different values of K_1

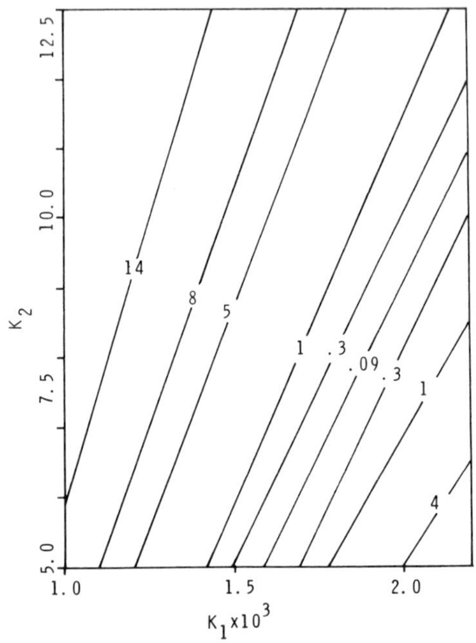

Fig. 3. RSS map in parameter space K_1 and K_2.

and K_2 was used. The corresponding error map is shown in Fig. 3 which comprises kinetic data from 0 to 93% conversion, for the rate experiment conducted at 0·002 ionic strength. The contour map clearly defines a valley, indicating that based on the minimum residual sum of squares criteria, an infinite number of solutions for K_1 and K_2 exists along the minimum error ridge. Only a few options exist in the calculation of K_1 and K_2:

(i) K_1 can be independently calculated from measured quantities and K_2 can be evaluated with the aid of K_1 from the grid.

(ii) The rate equation can be written in terms of K_1 and K_2 as:

$$R_p = K_1 \left[1 + K_2 \left(\frac{d_b}{d} \right)^2 \right]^{-1/2}$$

and using the calculated K_1, K_2 is accessible.

(iii) Using the relationship between K_1 and K_2, as obtained in the RSS map, a second relationship is given, which then, with the first, gives an independent estimation of K_1 and K_2.

The most interesting aspect of this analysis is, of course, the possibility of comparing the theoretical K_1 with the experimental K_1, calculated in (iii). In order to determine the functional relationship between K_1 and K_2 along the RSS minimum, it is desirable to reconstruct the grid in a different manner. The K_1 parameters are held constant and the RSS is plotted against K_2. At the minimum, the corresponding K_2 yields the best estimator. This is shown in Fig. 4. From such plots, K_1 and K_2 pairs are obtained, defining a functional relationship between the parameters. Plotting this K_1 versus K_2, a linear curve is obtained as shown in Fig. 5. This linear equation can be expressed as:

$$K_1 = 1{\cdot}05 \times 10^{-3} + 1{\cdot}1 \times 10^{-4} K_2$$

and from the rate equation, the following relationship is obtained:

$$6{\cdot}40 \times 10^{-4} = K_1(1 + K_2)^{-1/2}$$

This, then, gives two equations in two parameters, which yield a quadratic equation with one root, equal to $1{\cdot}87 \times 10^{-3}$. This value is for K_1.

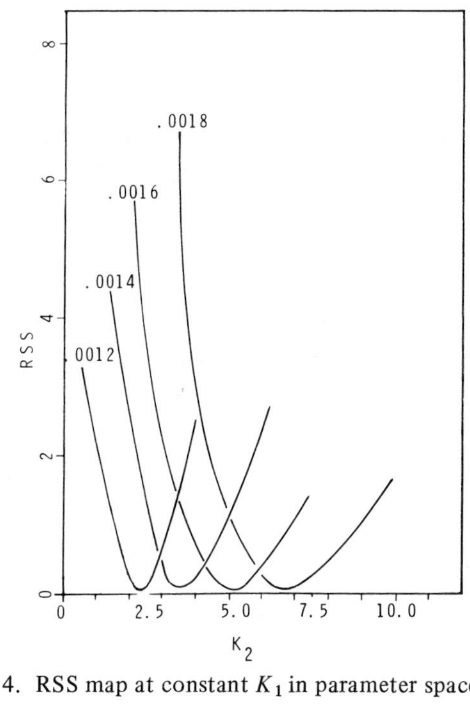

Fig. 4. RSS map at constant K_1 in parameter space K_2.

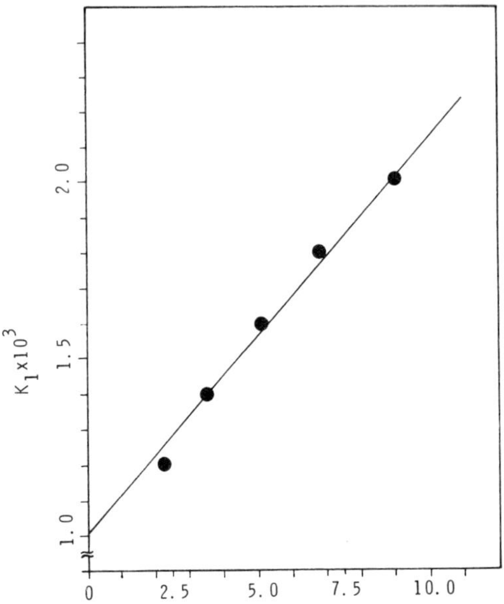

Fig. 5. The relationship of K_1 to K_2 along the RSS ridge for a single experimental run.

The parameter K_1 can be calculated independently from rate constants available from the literature, from measured solubilities of monomer in particle, and from particle size counts and initiation rate measurements conducted in this study. This theoretical K_1 value was calculated to be $1{\cdot}94 \times 10^{-3}$.

Now it is possible to take this theoretical K_1 value and calculate K_2 from the RSS relationship of K_1 versus K_2, or from the rate relationship. The former gives K_2 a value of 8·10, while the latter gives it a value of 8·18. In order to visually observe the fit between the experimental points and the theoretical curve obtained from $K_1 = 0{\cdot}00194$, and $K_2 = 8{\cdot}18$, Fig. 6 shows both on the same graph.

The agreement between the K_1 parameter obtained by statistical means, $1{\cdot}86 \times 10^{-3}$, and the one determined from literature data and experimentally measurable variables, $1{\cdot}94 \times 10^{-3}$, is rather good. Similarly, the K_2 values calculated either by the use of the RSS ridge equation, 8·10, or from the rate equation, 8·18, yields good agreement. The good fit to experimental data justifies the simplified approach for determining K_2, i.e. first evaluate K_1 from literature and experimental data and use the rate equation to determine K_2.

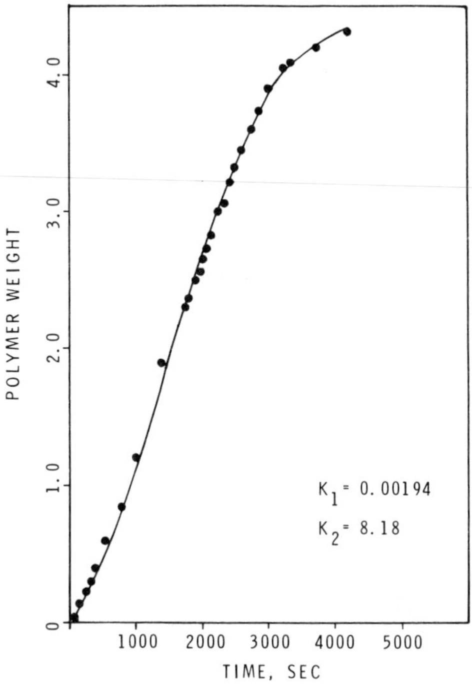

Fig. 6. The theoretical fit to experimental data on a conversion time scale.

The rate constant for oligomeric radical capture is a part of the parameter K_2. Using the measured and literature values for particle size, the monomer solubility in particles and in water, and the chain transfer rate constant, the parameter K_2 is reduced to the following combination of rate constants:

$$\frac{k_{10.2}W}{k_8 k_4^{\circ}}$$

In this expression only W is a function of ionic strength, thus a logarithmic plot against ionic strength would result in the separation of variables W and the constant $k_{10.2}/k_8 k_4^{\circ}$.[13]

In Table VIII the parameters K_1 and K_2 with the stability factor W and the capture rate constant are shown as a function of ionic strength. The stability factor as a function of ionic strength is shown in Fig. 7. The decreasing stability with increasing ionic strength is attributed to the effect of the shrinking electric double layer. The increase in stability at higher ionic strengths, like in soap stabilized oil in water emulsions,[17] is explained by the

TABLE VIII

Ionic strength dependence of stability factor and rate constant for the sweep-up of radicals, at 60·2°C

Ionic strength	*pH*	$K_1 \times 10^3$	k_2	*W*	$k_4 \times 10^{-8}$
0·002	3·55	1·940	8·188	38·9	3·62
0·0036[a]	—	—	—	8·9[a]	15·80[a]
0·008	4·65	0·853	0·799	4·78	29·5
0·019	4·40	1·053	1·408	6·32	22·2
0·015	5·27	1·009	1·275	4·78	29·5
0·045	6·36	1·009	2·530	10·0	14·1
0·049	4·52	1·103	3·23	13·5	10·4
0·075	4·77	1·020	3·21	12·9	10·9

[a] Data from Litt *et al.*[4]

Notes: K_1 in mole/liter s

k_4 in liter/mole μm^2 s

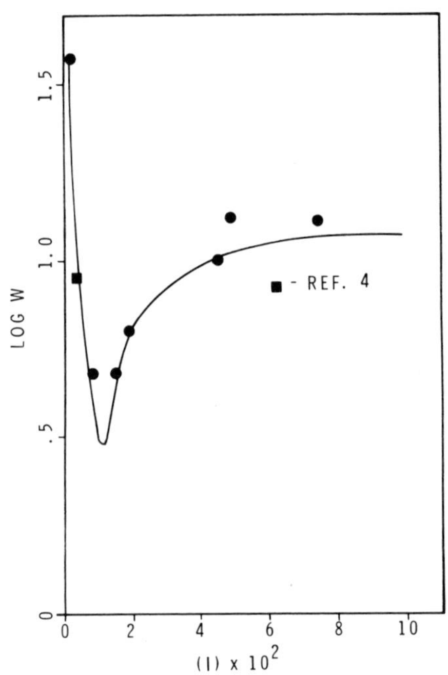

Fig. 7. Ionic strength (*I*) dependence of the stability factor at 60·2°C.

increased adsorption of surfactant from the water phase onto the particle surface.

The minimum in stability occurs at an ionic strength of 0·013 M. The electrolyte coefficient of stability in the region of the shrinking double layer was found to be −2·5. This value is very much in line with those values found for polystyrene latexes.[12] Coagulation studies on sodium lauryl sulfate solubilized poly(vinyl acetate) yielded a critical coagulation concentration of $5{\cdot}3 \times 10^{-3}$ M K_2SO_4, which is 0·015 M ionic strength units.[12] The shrinkage of the electrical double layer for sodium lauryl sulfate stabilized poly(vinyl acetate) reaches a critical value of about 0·014 ± 0·002 M ionic strength. Below this concentration the electrolyte effect on the double layer strongly influences the stability factor and the rate of radical capture.

Between ionic strengths 0·014 and 0·03, the stability factor increases with a corresponding lowering in the capture rate constants, which is due to the formation of an adsorbed, condensed surfactant film at the particle interface. Above 0·03 ionic strength, in the range of 0·03–0·08 M, no effect was found on W. It is reasonable to expect that the stability factor W derived for the particle oligomeric radical collisions applies to particle collisions with all solubilized oligomeric polymer, regardless of radical activity. Since termination is postulated to be in the aqueous phase and relatively few radicals enter the particles, there must be some finite concentration of solubilized oligomeric polymer. These oligomeric species would also be captured by the particles, effectively increasing the weight of the particles by water phase polymerized vinyl acetate. (Substantial water phase polymerization was shown before.[16]) Therefore, the particle volume should follow the increase trends in the stability factor. This is shown in Fig. 8. As the stability decreases, particle volume increases and as stability increases, particle volume decreases in the region of 0·002 to 0·03 ionic strength. Above 0·03 to 0·075 ionic strength, both the stability factor and particle volume remain essentially constant.

The rate dependence on initiator concentration is governed by ionic strength. Below an ionic strength of 0·014 M, the rate expression simplifies to

$$R_p = \frac{k_5}{k_6}\left[\frac{k_1}{k_{10\cdot 2}} k_4 d^2 k_8 [I][N_p][M_{aq}]\right]^{1/2}$$

because

$$K_2 = \frac{k_{10\cdot 2} k_6 [M_p]}{k_4 k_8 d^2 [M_{aq}]} \gg 1$$

which is due to the small k_4 and d values.

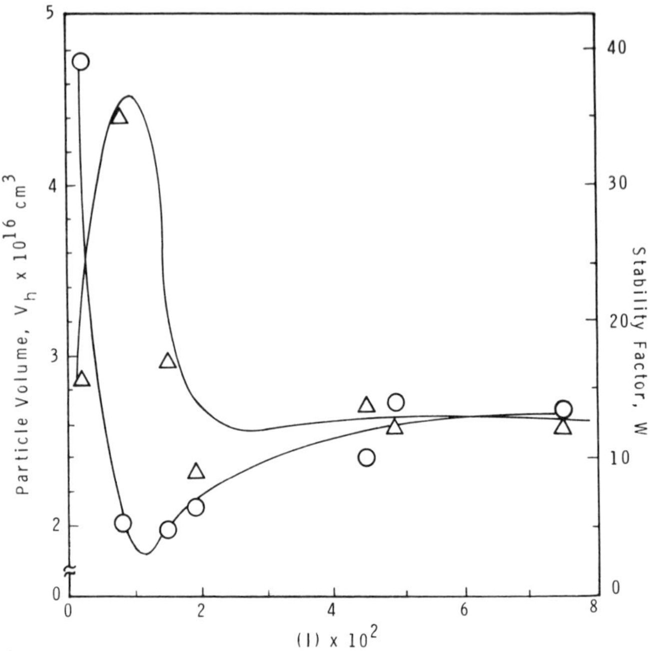

Fig. 8. Effect of ionic strength on particle volume, ○, and stability factor, △.

At low ionic strengths the dependencies can be approximated:

(i) $d \propto [I]^{0.13}$

(ii) $[N_p] \propto \dfrac{1}{d^3}$

(iii) $k_4 \propto [I]^{0.7}$

which in the simplified rate expression at low ionic strength yields

$$R_p \propto [I]^{0.8}$$

Patsiga for seeded polymerization found the same power dependence[16] and Zollars at low ionic strength found a 0·88 ± 0·14 power dependence.[6]

At higher than 0·014 M ionic strength when

$$K_2 \ll 1$$

the rate equation reduces to

$$R_p = \frac{k_5}{k_6}[(k_6 k_1 [I][M_p][N_p])]^{1/2}$$

This reaction rate dependence on the initiator concentration to the half power was reported by Zollars,[6] Nomura *et al.*,[2] and Gershberg.[3]

The pH and ionic strengths were impossible to vary independently. At a low ionic strength, the buffer capacity was low and at high ionic strength, instability was found, which limited the pH range. In order to explore the pH dependence of the rate constant for radical sweep-up, k_4, a contour diagram was prepared on a pH versus ionic strength grid. Contour graphing is a well known statistical tool to represent complex relationships depicting some unknown model. The method is similar to that employed in topographical and meteorological studies. This contour graph is shown in Fig. 9. At a constant ionic strength the indication is that k_4 increases with increasing pH.

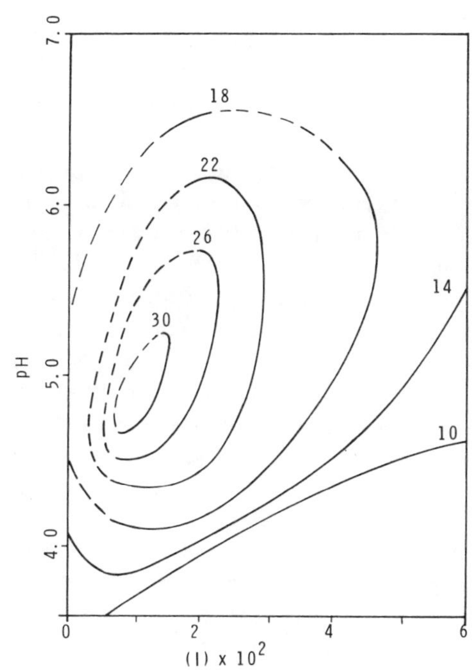

Fig. 9. Contour diagram for the rate constant for radical sweep-up, k_4 in ionic strength and pH space at 60·2°C.

Viewing this tendency in terms of the stability factor, W, it means that W is decreasing with an increase in pH, while ionic strength remains a constant. From a surfactant equilibrium adsorption point of view, this would mean that the amount of surfactant adsorbed increases with decreasing pH.

Such is the case found for sodium dodecyl sulfate adsorption on aluminum oxide,[18] and some other metal oxide substrates. The mechanism of adsorption on these polar substrates is understood in terms of pH dependent polar adsorption sites. It was indicated for aluminum oxide that, with decreasing pH, cationic adsorption sites increased. Chemisorption of anionic surfactant onto these sites occurred, due to the electrostatic attraction forces. A second layer of surfactant, then, may adsorb onto the first by van der Waals forces.

There is some objection that such is the case for dodecyl sulfate adsorption on poly(vinyl acetate) particles in the present study. The pH dependent cationic sites could occur on the particles through the formation of oxonium ions from surface hydroxyls, which are present due to hydrolysis. It is doubtful that even at the lowest pH, 3·4, hydrogen ion concentration is high enough for the reaction to occur.

An alternative hypothesis to explain the increased pH dependent destabilization of particles may be advanced, as follows. In the Appendix it is shown that polyvinyl alcohol adsorbs acetate ions. Provided that hydroxyl sites on the surface of the particle are exposed, acetate ions could associate with them. When the acetate adsorbs, the electrical interaction between the surfactants adsorbed on neighboring sites would force the desorption of surfactant. In such a case, a reduction in the surface potential would follow. This reduction would result from a greater counter-ion penetration of the complexed acetate. This would mean a lower stability. At low pH, however, the hydrolysis-produced acetic acid exists almost entirely as the associated species. As the pH increases, more and more acetate ions become available to complex with the hydroxyl of the particle, resulting in a lower stability.

As the ionic strength increases, however, the particle surface is more densely covered with surfactant and less hydroxyl is exposed for complexing. Hence, at higher ionic strengths the pH effect in lowering stability is less pronounced.

The acetone effect on the rate of polymerization is about 10–15% in the range of 50–3000 ppm, with a maximum in rate occurring at about 300 ppm. There is no hypothesis to explain the behavior even though at low ppm acetone content surface effect is suggested,[13] while at high ppm acetone content some destabilization of particles may occur.

The decomposition of persulfate in the polymerizing latex was determined *in situ* polarographically. Ionic strength had no effect on the decomposition rate. The activation energy obtained for the hydrogen ion concentration independent rate constant of 33·2 kcal/mole compares well with the 33·5 kcal/mole obtained in water at an alkaline pH of 13.[19] The hydrogen ion catalyzed decomposition rate yielded 8·5 kcal/mole, while that in the literature is 26·0 kcal/mole. There is no good hypothesis at present to explain the large discrepancy for the hydrogen catalyzed activation energy obtained in latex versus water.

APPENDIX

SODIUM ACETATE ADSORPTION INTO POLYVINYL ALCOHOL

Polyvinyl alcohol is known to associate with multifunctionally charged dye molecules,[20] such as congo-red, to form, at high concentration, a gelled network. The reaction has been proposed to occur between the polyvinyl alcohol hydroxyl and the sulfonate group of the dye. At lower concentrations the increment in viscosity was attributed to an increase in hydrodynamic end-to-end distance of the polyvinyl alcohol, which was due to the repulsive interaction of associated charged dye molecules on the polymer backbone.

From the following viscometric and pH studies it will be shown that similar to the dye reaction, sodium acetate increases polyvinyl alcohol viscosity by, supposedly, chemisorption onto the hydroxyl group. It was found that the pH of a 1% polyvinyl alcohol solution varied in the presence of various amounts of sodium acetate. The pH markedly differed from what one would expect theoretically. This is shown in Table AI.

TABLE AI
Theoretical and measured pH of sodium acetate in the presence of polyvinyl alcohol

NaAc M	*pH Measured*	*pH Calculated*
	7·1	—
0·1	7·4	8·88
0·5	7·9	9·23
1·0	8·1	9·38

The theoretical aspects are given at the end of this section. It can be seen that the chemisorbed amount of acetate can be calculated as the difference between measured and initial acetate ion concentration, which in turn is related to pH. Thus, the discrepancy in pH noted before is due to the acetate uptake of the polyvinyl alcohol.

Carrying out the viscosity experiments with a conventional Ostwald type of viscometer on a commercial 88% hydrolyzed polyvinyl alcohol of number average molecular weight $1{\cdot}8 \times 10^5$,† the results shown in Table AII were obtained.

TABLE AII
Polyvinyl alcohol solution viscosities in the presence of sodium acetate

Concentration PVOH g/100 ml	*Concentration NaAc* M	*Viscosity*		
		Specific	*Reduced*	*Intrinsic*
0·4	0·05	0·384	0·960	0·81
0·8	0·05	0·884	1·106	
1·0	0·05	1·203	1·203	0·81
0·4	0·10	0·404	1·010	
0·8	0·10	0·895	1·119	
1·0	0·10	1·209	1·209	0·94
0·5	0·50	0·624	1·248	
0·8	0·50	1·018	1·272	1·10
1·0	0·50	1·354	1·354	

The intrinsic viscosities of charged polymers have been correlated with hydrodynamic end-to-end distance $\mathfrak{h}$ by the following relationship:[21]

$$[\eta] = \frac{\pi \mathfrak{b}\, \mathfrak{h}^2 N_A}{32 \times 10^{13}}$$

where N_A = Avogadro's number, $\mathfrak{b}$ = length of monomeric unit on molecular axis. From this relationship it can be seen that the ratio of two intrinsic viscosities would reveal the ratio of the corresponding mean square hydrodynamic end-to-end distances. Thus, if the intrinsic viscosity of the uncharged species was available, it could serve as a basis for comparison.

† Elvanol 52-22 (DuPont's Trade Mark).

For this reason the intrinsic viscosity versus square root of ionic concentration relationship was linearized by the well-known empirical method of Davies[22] to obtain the following relationship:

$$\log([\eta] + 1{\cdot}11) = \log[\eta]_0 - m'(C_p)^{1/2}$$

This is a linearized function and from the intercept, $[\eta]_0$ is found to be 0·74.

As a function of sodium acetate ion concentration, the ratio of the root mean square end-to-end distance with respect to the uncharged polymer is available and is shown in Table AIII.

TABLE AIII
Hydrodynamic variables for polyvinyl alcohol as a function of sodium acetate concentration

Concentration NaAc M	*Intrinsic viscosity*	$h^2/h_0^2{}^{1/2}$
0	0·74	1·0
0·05	0·81	1·05
0·10	0·94	1·13
0·50	1·10	1·22

It has been shown that in the presence of polyvinyl alcohol, acetate ion concentration decreases in the aqueous phase. Polyvinyl alcohol is supposed to chemisorb the acetate ion. This is documented by the expected increase in intrinsic viscosity of the polymer as the charged adsorbates interact on neighboring sites, extending the root mean square hydrodynamic end-to-end distance.

Theoretical

The following equilibrium relationships are used or derived to determine the amount of acetate ions adsorbed on the polyvinyl alcohol:

$$H^+ + {}^-OH \leftrightharpoons H_2O, \quad K_w = 10^{-14} \tag{A1}$$

$$H^+ + {}^-Ac \leftrightharpoons HAc, \quad K_a = 1{\cdot}75 \times 10^{-5} \tag{A2}$$

Since any hydrogen ion for reaction (A2) to take place must come from reaction (A1)

$${}^-OH = HAc \tag{A3}$$

and therefore

$$H^+ = K_a \frac{^-OH}{^-Ac} \tag{A4}$$

Equation (A4) can also be written with the aid of eqns. (A1) and (A3) as:

$$K_w/^-OH = K_a\, ^-OH/^-Ac \tag{A5}$$

which, when solved for acetate ion concentration, yields:

$$^-Ac = K_a K_w {}^+H^{-2} = 1{\cdot}75 \times 10^{-19} \times 10^2\ pH$$

The amount of adsorbed acetate from the material balance is:

$$^-Ac_{input} - 1{\cdot}75 \times 10^{-19} \times 10^2\ pH = {}^-Ac_{adsorbed}$$

which can be calculated.

REFERENCES

1. N. Friis and L. Nyhagen, *J. Appl. Polym. Sci.*, **17**, 2311 (1973).
2. M. Nomura, M. Harada, K. Nakagawara, W. Eguchi and S. Nagata, *J. Chem. Eng. Jpn.*, **4**, 160 (1971).
3. D. Gershberg, *AIChE Symposium Series*, **3**, 4 (1965).
4. M. H. Litt, R. Patsiga and V. T. Stannett, *J. Polym. Sci., A1*, **8**, 3607 (1970).
5. P. Harriot, *J. Polym. Sci., A1*, **9**, 1153 (1971).
6. H. L. Zollars, *J. Appl. Polym. Sci.*, **24**, 1353 (1979).
7. W. V. Smith and R. H. Ewart, *J. Chem. Phys.*, **16**, 592 (1948).
8. V. T. Stannett, A. Klein and M. H. Litt, *Br. Polym. J.*, **7**, 139 (1975).
9. J. Ugelstad, *Polym. Colloids Preprints,* Nato Adv. Study Inst., University of Trondheim, 1975.
10. A. Klein, C. H. Kuist and V. T. Stannett, *J. Polym. Sci.*, **11**, 2111 (1973).
11. R. M. Fitch, *Polym. Colloids Preprints,* Nato Adv. Study Inst., University of Trondheim, 1975.
12. A. Klein and V. T. Stannett, *J. Coll. and Int. Sci.*, **77**, 76 (1980).
13. A. Klein, Ph.D. Thesis (Chemical Engineering), North Carolina State University at Raleigh, 1971.
14. I. M. Kolthoff, L. S. Guss, D. R. May and A. I. Medalia, *J. Polym. Sci.*, **1**, 340 (1946).
15. F. H. De La Court and H. Vogt, *J. Oil Col. Chem. Assoc.*, **52**, 587 (1969).
16. R. A. Patsiga, Ph.D. Thesis, State University College of Forestry at Syracuse University, 1962.

17. L. Osipow, S. Birsan and F. D. Snell, *J. Amer. Oil Chemists' Soc.*, **34**, 34 (1957).
18. K. T. Shinoda, B. Nakagawa, B. Tamamushi and T. Isemura, *Colloidal Surfactants,* Academic Press, New York, 1961, p. 224.
19. I. M. Kolthoff and I. K. Miller, *J. Amer. Chem. Soc.*, **73**, 3055 (1951).
20. C. Dittmar and W. J. Priest, *J. Polym. Sci.*, **18**, 275 (1955).
21. A. Katchalski, D. Kwizle and W. Kahn, *J. Polym. Sci.*, **5**, 283 (1950).
22. D. W. Davies, *Nomography and Empirical Equations,* Reinhold Publishing Co., New York, 1955.

6

The Reinvestigation of Vinyl Acetate Emulsion Polymerization (I) – The Rate of Polymerization

K. H. S. CHANG, M. H. LITT and M. NOMURA†

Department of Macromolecular Science, Case Western Reserve University, Ohio, USA

SUMMARY

The kinetics of the seeded emulsion polymerization of vinyl acetate were investigated thoroughly. The variation of the polymerization rate with changes in particle concentration, type of emulsifier, emulsifier concentration, persulfate concentration, ionic strength, and monomer volume were determined. The rate of polymerization is dependent on the initiator concentration to the 0·6 power, the particle concentration to the 0·12 power, and the vinyl acetate volume to the 0·39 power; however, it is independent of the type of emulsifier, emulsifier concentration, and ionic strength in the emulsion. In all cases, the rate of the polymerization is almost independent of monomer concentration in the particles until 85 or 90% conversion.

The results were rationalized by the following mechanism.

(a) The persulfate ion radical enters the particle where it either initiates or terminates the polymerization depending on whether the particle contains a radical or not. As only 1 to 2% of particles contain radicals, it usually initiates polymerization.

(b) The radical in a particle can chain transfer to monomer generating a monomer radical. It is believed that the kinetically important chain transfer is on vinyl hydrogen. This radical reinitiates relatively slowly and thus the radical can escape from the particle to become an aqueous radical.

(c) The aqueous monomer radical can do several things. (i) It can be swept up by a dead particle, (ii) it can react with initiator to generate a sulfate ion

† On sabbatical leave from Fukui University, Fukui, Japan.

radical which will then enter a particle and initiate or terminate the polymerization, (iii) it can be swept up into a particle containing a radical. Since the time the monomer radical stays in a particle is relatively short, the radical may diffuse out of such a particle without terminating, though it usually terminates under our conditions of polymerization.

(d) The escape from particles and diffusion back in continues for the monomer radical until it either reinitiates or terminates.

The kinetic steps were justified by data from the literature and kinetic equations were derived and integrated. The equations were compared with the experimental data and shown to match very well over the whole range of experimental parameters. The integrated equation followed the course of polymerization exactly over the whole parameter range. The rate constant for reinitiation was calculated to be 29 liter/mole s and for termination to be $1{\cdot}1 \times 10^9$ *liter/mole s at 60°C.*

NOMENCLATURE

a	partition coefficient of monomer and monomer radicals between the particles and the aqueous phase
c	fractional conversion of polymer
d	particle diameter
D	diffusion constant of monomer radical in particles
D_a	diffusion constant of monomer radical in aqueous phase
D_0	diffusion constant of monomer radical in particles before disappearance of separate monomer phase
f	the efficiency of unimolecular persulfate decomposition
(I)	initiator concentration
(I_0)	initiator concentration at beginning of reaction
k_1	unimolecular decomposition constant of potassium persulfate
k_2	propagation rate constant of vinyl acetate
k_3	chain transfer constant of growing radical to monomer
k_4	reinitiation rate constant, chain transferred monomer radical adding to monomer
k_7	induced decomposition rate constant, attack by monomer radical on potassium persulfate in the aqueous phase
k_{80}	termination rate constant for monomer radical in a particle at or before 30% conversion

K_2	$= 2{\cdot}75 \times 10^{-9} k_7 (fk_1 k_3/k_4)^{1/2}$
K_3	rate of polymerization at 30% conversion
(M)	monomer concentration in particles
(M')	monomer concentration per liter H_2O during polymerization
$(M_1\cdot)$	concentration of particles containing initially chain transferred monomer radical
$(M\cdot_a)$	monomer radical concentration in aqueous phase
(M'_0)	initial monomer concentration/liter H_2O
(M_{a_0})	monomer concentration in aqueous phase before the disappearance of separate monomer phase
(M_a)	monomer concentration in aqueous phase
$(M\cdot_p)$	concentration of particles into which a monomer radical has diffused
$(M\cdot_p P\cdot)$	concentration of particles containing one growing radical and one monomer radical
N	Avogadro's number, $6{\cdot}02 \times 10^{23}$
$\bar{n}$	fraction of particles with growing radicals
(P)	concentration of particles/liter H_2O
$(P\cdot)$	$= (P\bar{n})$ = concentration of particles with growing radicals/liter H_2O
P_e	probability of escape of monomer radical from particle $(= 1/(1 + k_4 Mad^2/4D))$
P_t	probability of termination of monomer radical with growing radical in a particle $(= 1/(1 + 2N\pi dD_0/3ak_{80}))$
Rp	rate of polymerization
V_{org}	volume of organic phase contained in the particles
V_{aq}	volume of aqueous phase

INTRODUCTION

The emulsion polymerization of vinyl acetate has been studied extensively in recent years.[1-8] It is clear from these results that it does not follow the kinetics derived by Smith and Ewart for styrene.[9] In addition, the experimental results differ widely in different papers, and are contradictory in many respects; however, no group of researchers used exactly the same experimental conditions.

There is some agreement about the following features of the reaction:

(a) The rate of polymerization is approximately zero order with respect to monomer concentration at least from 20 to 85% conversion.

(b) In unseeded polymerizations, the number of particles is roughly independent of conversion after 30% conversion.
(c) The polymerization rate depends on the particle concentration to about 0·2 power.
(d) The dependence of the polymerization rate on the emulsifier concentration is small with a maximum of 0·25 power.
(e) The molecular weights are independent of all variables and mainly depend on chain transfer to monomer.

The major disagreements are on the order of dependence of the rate on initiator concentration which ranges from 0·5 to 1·0 and, in unseeded polymerizations, the dependence of the particle concentration on the initiator concentration, which varies from zero to 1·2. These differences appear to be genuine since the induction periods are similar and the rates of comparable order of magnitude. Table I is a summary by Stannett *et al.*[10] of different results reported by various researchers. Only those dealing with the potassium persulfate and sodium lauryl sulfate system are listed in the table. The polymerization conditions varied significantly for different studies. The first significant variable was temperature, 50°C for Friis and Nyhagen[1] and Nomura *et al.*[2]; and 60°C for Patsiga,[3] Litt *et al.*[4] and Dunn *et al.*[7,8] The second variable, the monomer–water phase ratio, was approximately the same for all above authors except Dunn *et al.*[7,8]

TABLE I
Summary of data on vinyl acetate emulsion polymerization as reported in the literature

Order of rate (Rp) with respect to:			*Order of number of particles (P) with respect to:*				
$K_2S_2O_8$	*SLS*†	*P*	$K_2S_2O_8$	*SLS*	*Temperature (°C)*	*Author (et al.)*	*Reference*
0·5	0·15	—	~0	0·9	50	Nomura	2a, b
0·6	0·25	—	—	0·2	50	Gershberg	5
0·7–0·8	~0	—	—	—	50	Breitenbach	6
0·64	0·1	—	—	—	60	Dunn 1965‡	7
1·0	—	—	~1·0	—	60	Dunn 1970‡	8
0·5	~0	0·25	~0	0·50	50	Friis	1
0·8–1·0	~0	0·2	1·2	0·2	60	Patsiga	3, 4

† SLS = sodium lauryl sulfate.
‡ The work of Dunn *et al.* was carried out with no separate monomer phase.

A number of models have been developed[1,2,4,5,7,11,12] but they either do not explain all the results or postulate improbable kinetic steps. Of these various models proposed to interpret the kinetics and other features of vinyl acetate polymerization those of Friis,[1] Litt *et al.*[4] and Nomura *et al.*[2] appear to be the most reasonable. However, none are perfect; in many respects they cannot explain all the observed relationships, and the models contradict each other.

The LPS (Litt–Patsiga–Stannett) model postulated reinitiation in the monomer poor aqueous phase, and kinetically important aqueous polymerization with high dependence on ionic strength. Other investigations[10] have found very low dependence on ionic strength.

The major difference between the LPS model[4] and others[1,2] is that it postulates solubilized PVAc (poly(vinyl acetate)) radicals in the aqueous phase stabilized by sodium lauryl sulfate (SLS). (One of the rationales for this work was to test this hypothesis by comparing three emulsifiers: SLS, one which did not stabilize the polymer, and one which solubilized it much more than SLS. If the kinetics were affected by the soap with all else held constant, the aqueous phase is kinetically important.)

Friis' model[1] does not explain the frequent occurrence of a greater than 0·5 dependence of the rate on the persulfate concentration. In Nomura's model,[2] there is a discrepancy between the observed and the calculated average number of radicals per particle, $\bar{n}$, after 35% conversion. He explained the discrepancy between the observed and calculated value of $\bar{n}$ by the so-called gel-effect. Also his model postulated that the chain transferred monomer radical reinitiated as fast as the propagating polymer radical, which is not necessarily true, and would not give the observed kinetics.

Chain transfer is an important phenomenon in vinyl acetate emulsion polymerization. Most reports concluded that most or all of the monomer transfer occurred at the acetyl hydrogens[13,14] and this has been generally accepted. However, Wheeler *et al.*,[15] by analyzing for unsaturation, calculated a ratio of transfer on vinyl hydrogen (α or β hydrogens) to transfer on acetyl hydrogen of about 0·33 to 0·50. Howard[16] studied the transfer to solvents, monomer, and polymer and found a vinyl/acetyl chain transfer ratio which agreed with Wheeler.

Litt *et al.*[4] have argued reasonably that most of the acetyl chain transferred monomer radicals cyclize to form a butyrolactonyl radical which propagates as fast or faster than the corresponding acetyl monomer radical

$$\cdot CH_2{-}\overset{\overset{\displaystyle O}{\|}}{C}{-}O{-}CH{=}CH_2$$

or growing radical. If this reinitiates rapidly compared to diffusion out of particles,[1] such a chain transfer could not affect the kinetics. This will be discussed in more detail later. It is possible that if chain transfer is kinetically important, which everyone believes, transfer on the vinyl hydrogens of vinyl acetate is the major, kinetically significant chain transfer step.

The present chapters (6, 7 and 8) investigate in detail the parameters which affect vinyl acetate emulsion polymerization. In this chapter, the rate of the polymerization will be discussed and a new kinetic model developed.

EXPERIMENTAL

Materials

Monomer

Vinyl acetate was obtained from Eastman Kodak Company and purified by distillation through a Perkin Elmer spinning band column with 45 theoretical plates. The first and the last 25% of the distillate was discarded for all the distillations. It was then stored at 5°C.

Emulsifiers

Three emulsifiers with widely different solubilizing power for poly(vinyl acetate) were chosen for kinetic studies. A highly purified toothpaste grade of sodium lauryl sulfate (Stepanol WA-100) was obtained from Stepan Chemical Company and was used directly. A purified grade of sodium lauryl ether sulfate (~30 ethylene oxides, Polystep B-19), 28 wt% concentration in aqueous solution, obtained from Stepan Chemical Company was used directly from the bottle. The third emulsifier was sodium tridecyl sulfate aqueous solution, 25 wt% emulsifier, which was obtained from Alcolac Corporation. It was used directly from the bottle, and also after purification.

It was purified as follows.

Fifty grams of sodium tridecyl sulfate solution was dried under vacuum using a rotary evaporator. Twelve grams of solid was obtained. It was then extracted with ether for 20 minutes using a Soxhlet extraction apparatus. Four grams of a white residue was obtained which was found to be reasonably pure material. Seven grams of a yellow extract was obtained after evaporating the ether. The purified sodium tridecyl sulfate and the extract were compared.

Initiator

The initiator, potassium persulfate, was Fisher certified reagent grade and was used directly from the bottle. Potassium sulfate which is added to the polymerization system to control the ionic strength was MC/B reagent grade and was also used directly.

Nitrogen

The nirogen used was very pure nitrogen, obtained from Matheson Gas Products (research purity, purity $>$ 99·9995%, $O_2 < 1$ ppm).

Poly(vinyl acetate) solubilization measurement

Before choosing appropriate emulsifiers for kinetic study, it was necessary to study the degree of solubilization of poly(vinyl acetate) in different emulsifiers. Eighteen emulsifiers were used in the study, as received from various companies. The suppliers were Alcolac (Abex, Sipex), Stepan Chemical Co. (Stepanol, Polystep) and American Cyanamid Co. (Aerosol).

The degree of solubilization of poly(vinyl acetate) in soap was estimated by the increase in the reduced specific viscosity of a 1% latex polymer in 3% soap solution, following the procedure of Isaacs and Edelhauser.[17] Viscosity measurements on the above system were made in an Ubbelohde viscometer in a water bath at 30·00 ± 0·02°C. Reduced specific viscosities were calculated relative to the pure surfactant solution.

Preparation of seed latex

The apparatus used is shown in Fig. 1. The polymerization was conducted at 60 ± 0·05°C under nitrogen atmosphere (Matheson nitrogen, purity $>$ 99·9995%). Water from a constant temperature bath was passed through the water jacket of the reaction vessel and served as the heating medium during the polymerization. Before heating, the water and vinyl acetate in the vessel were flushed for 30 to 40 minutes at room temperature with the pure nitrogen. During the flushing, stopcock C was closed and stopcock D was open (Fig. 1). While the water/vinyl acetate mixture was reaching 60°C, the initiator solution in the initiator feeder, G, was flushed with nitrogen for five minutes by opening stockcocks E, F and N. Stopcock N was then closed while the initiator feeder G was kept flushing for another three minutes. Stopcocks F and E were closed successively after the feeder G was flushed. Stopcock N was opened to allow the residual nitrogen in the side arm of feeder G to flow above the initiator solution. The initiator solution then flowed into the reactor when stopcock F was opened, starting the polymerization. The polymerization was finished in one to five hours. A typical recipe is given in Table II.

The seed latex (80 g) was then washed with water (300 g) through a Pellicon filter with a nominal molecular weight limit of 10 000 to remove salts

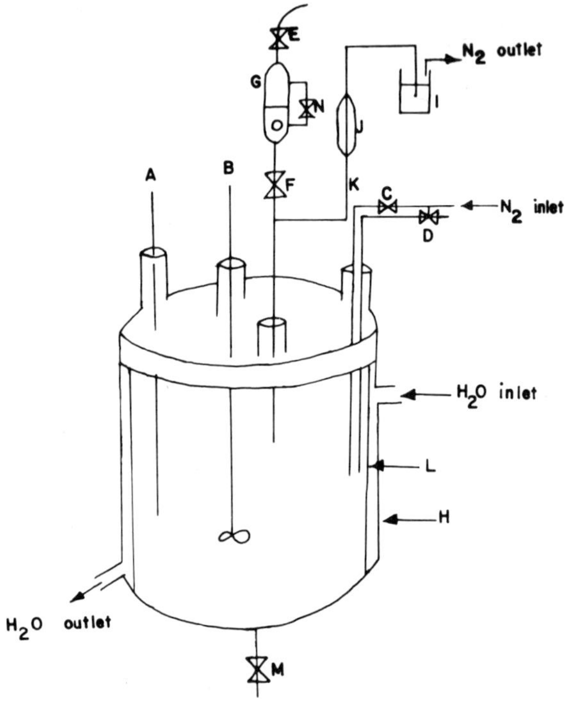

Fig. 1. Diagram of experimental apparatus: A, thermometer; B, stirrer; C, D, control cocks for N_2 inlet; E, control cock for vacuum; F, N, cocks controlling the initiator feeding; G, feeder for initiator; H, reaction vessel; I, pressure regulator; J, reflux condenser; K, adaptor; L, water jacket; M, sampling cock; O, initiator solution.

TABLE II
Typical recipe for the preparation of seed

Water: 600 g
Initiator solution: 10 g (0·4205 g $K_2S_2O_8$/50 g H_2O)
Soap solution (Polystep B-19): Sodium lauryl ether sulfate ($C_{12}H_{25}(OCH_2CH_2)_{20}OSO_3Na$) 12 g (4·2 g active)
Monomer: 120 g vinyl acetate

and the residual emulsifer in the aqueous phase. The apparatus for the filtration of the seed consisted of a Millipore 47 mm diameter stirred cell filter holder attached to a half water filled 600 ml liquid filling system which was pressurized from a nitrogen tank. The rate of filtration was controlled by the nitrogen pressure; usually about 2 bar was used which gave a steady state flow rate of 0·29 ml/min. The washed seed latex was then stored at 5°C. Two seed latexes were made; one had a solids content of 17·0% and the other 19·4% and their number average diameters were measured as 0·045 and 0·046 μm respectively by quasi-elastic laser light scattering.[18]

Apparatus for measuring polymerization rate

The polymerizations were conducted in a dilatometer, Fig. 2, modified from that designed by Corrin.[19] The reactor used was made from a 250 ml volu-

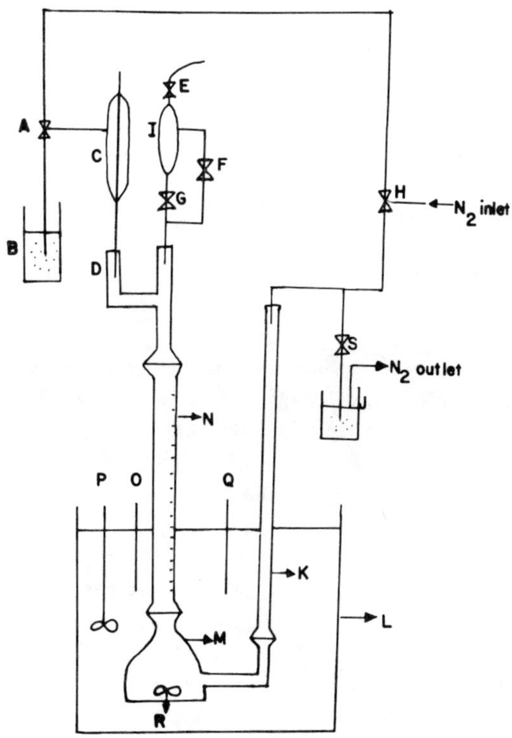

Fig. 2. Dilatometer: A, H, three-way cocks controlling nitrogen flow; B, J, pressure regulators; C, reflux condenser; D, adaptor; E, control cock for vacuum; F, G, cocks controlling initiator feeding; I, initiator feeder; K, precision bore tube; L, cylindrical jar used as water bath; M, reaction vessel; N, calibrated tubing (25 cc) with O-ring joints; O, heater; P, stirrer; Q, temperature regulator; R, magnetic stirrer; S, cock regulating N_2 flow.

metric flask with three indentations at 120° and contained a magnetic stirrer. A precision bore tube with an inside diameter of 6·0 mm was connected at one side and a precision, graduated 25 ml pipet with an inside diameter of 12·0 mm which could be read reproducibly to 0·0125 ml was modified to connect at the center of the flask with O-ring joints.

Polymerization procedure

All the polymerizations except one were conducted at 60 ± 0·05°C in a water bath under high purity nitrogen. Before the water bath was heated, the reactants in the vessel were flushed with Matheson research purity nitrogen for 30 to 40 minutes to eliminate oxygen. While flushing, the nitrogen went through a three-way stopcock, H, to the precision tube, K, bubbled through the liquid and was exhausted at A after passing through the adapter, D, and condenser, C. Then the aqueous initiator solution in the feeder, I, was flushed with nitrogen for five minutes by opening G and F and E and evacuating slightly. E, F and G were then closed again. F and G were opened to let the initiator solution flow into the reactor. After three to five minutes, the polymerization started.

Stirring speeds were kept as constant as possible for each run in the range of 120 to 150 rpm. The monomer conversion was measured by the volume contraction of the latex as the monomer polymerized. Nitrogen was directed from H through A until the pressure forced the latex level to a convenient reference point on the graduated column, N. At the same time latex was forced up the side arm (precision tube), the levels of latex in both tubes were measured using a travelling telescope. There is a drop of the latex level in the precision tube with polymerization. The total volume contraction was about 10–16 cc depending on the organic volume used. As this was greater than the precision tube volume, different reference points were chosen on the calibrated tube, N, as polymerization progressed. Conversion was calculated from volume contraction using a factor of 23·2 cm^3/mole.[3,4] The direction of nitrogen flow was reversed when measurements were not being made.

Most of the polymerization systems were based on a recipe which gave a latex with 25 wt% of polymer. The typical recipe used is shown in Table III. Conversions could be measured as often as once a minute.

The measurement of particle number

Particle sizes were measured by quasi-elastic laser light scattering.[18] The number of the particles can be calculated using the following expression:

$$(P) = \frac{w_0}{\pi d^3 \rho / 6} \tag{1}$$

TABLE III
A typical recipe for the seeded emulsion polymerization of vinyl acetate

H_2O: 142 g
Seed: 52 g (7·54 g polymer)
Initiator solution ($K_2S_2O_8$): 10 g (0·1335 g/50 g H_2O)
Soap solution (Polystep B-19): 1·8 g (0·5 g active)
Vinyl acetate: 58 g

where w_0 = weight of polymer per cc water in the latex, d = number average diameter of particles, ρ = density of polymer, (P) = number of particles per cc water. The number of the particles was checked before and after the polymerization. If new particles were generated, they were taken into account in the kinetic study.

RESULTS

All the results in this paper are concerned only with the seeded emulsion polymerization of vinyl acetate. This is because the number of particles in the latex can be well controlled by using seed, thus simplifying the analysis. The mechanism of particle formation was not considered here.

To study the effect of each parameter on the kinetics, experiments were run varying one parameter while the others were kept constant.

The solubilization of poly(vinyl acetate) by different emulsifiers

In order to see whether the kinetics are affected by differing aqueous solubilization of poly(vinyl acetate) radicals by different soaps, soaps of different solubilizing ability must be used. Eighteen different emulsifiers were chosen and solubilization measured. The degree of solubilization of poly(vinyl acetate) could be estimated from the specific viscosity of the solution. The higher the specific viscosity of the solution, the higher the degree of solubilization. Table IV lists the results of these measurements.

As can be seen, the mixture of poly(vinyl acetate) with Sipex TDS had a very high specific viscosity, which implies that poly(vinyl acetate) was highly solubilized by this soap. In fact this mixture was clear, while the others were cloudy. Conversely, the mixture of poly(vinyl acetate) with most other soaps except SLS showed low specific viscosity, indicating that poly(vinyl acetate) is not solubilized.

TABLE IV
The specific viscosities of poly(vinyl acetate) emulsifier aqueous mixtures (1% latex solids)

Trade name of emulsifiers	*Structure of emulsifier from manufacturer's data*	*Specific viscosity (η_{sp}) of the solution dl/g*
Sipex TDS	$C_{13}H_{27}OSO_3Na$	57·3
Stepanol WA-100	$C_{12}H_{25}OSO_3Na$(SLS)	4·70
Polystep B-19	$C_{12}H_{25}(OCH_2CH_2)_{30}OSO_3Na$	0·027
ABEX-18S	$C_{12}H_{25}(OCH_2CH_2)_{20}OSO_3Na$	0·030
Sipex EST-30	$C_{13}H_{27}(OCH_2CH_2)_nOSO_3Na$	0·063
Sipex EST-75	$C_{13}H_{27}(OCH_2CH_2)_mOSO_3Na$	0·042
Sipex BOS	$C_4H_9CH(C_2H_5)CH_2OSO_3Na$	0·049
Sipex WD	$C_{12}H_{25}OSO_3Na$(SLS)	4·66
Sipex ES	Sodium ether sulfate	0·048
Sipex JKB	Anionic surfactant	0·067
ABEX-26S	$C_{12}H_{25}(OCH_2CH_2)_{20}OSO_3Na$ + $C_{12}H_{25}(OCH_2CH_2)_{20}OH$	0·063
Polystep A-18	Sodium alpha olefin sulfonate	1·34
Polystep B-10	Sodium lauryl ether sulfate	0·055
Polystep B-16	Sodium salt of a fatty ether sulfate	2·18
Aerosol OT-100	Sodium dioctyl sulfosuccinate	0·042
Aerosol A-268	Disodium isodecyl sulfosuccinate	0·057
Aerosol A-102	Disodium ethoxylated alcohol	0·072
Aerosol 501	Sulfosuccinate surfactant	0·028

Dependence of polymerization rate on emulsifier type

The LPS[4] theory postulated that the mechanism of polymerization should change as emulsifier type changed. An emulsifier which solubilized poly(vinyl acetate) should cause the polymerization rate to drop compared to one which could not solubilize the polymer. Thus three emulsifiers were chosen for rate studies, Sipex TDS, Stepanol WA-100 and Polystep B-19 which had different solubilizing power for poly(vinyl acetate) latex. The polymerization rate of the run which had no added emulsifier was measured for comparison.

Figure 3 shows the comparison of the polymerization rates of different emulsifier systems. As can be seen the rates of polymerization, the slope of the linear portion of the curves, when zero soap, Polystep B-19 or Stepanol WA-100 were studied are essentially identical. However, when Sipex TDS

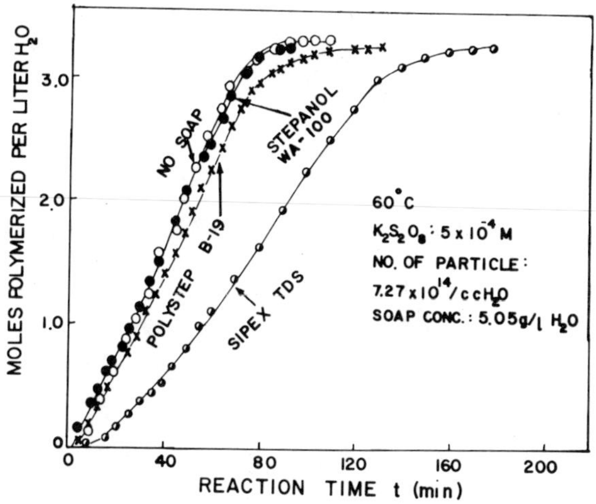

Fig. 3 The dependence of the polymerization rate on different type of soaps ($V_{org}/V_{aq} = 0{\cdot}33$, ionic strength ($\mu$) = 3×10^{-3}).

was used, the polymerization rate was considerably lower than that for any other kind of soap.

To find out why the rate was much smaller with Sipex TDS as the emulsifier, the soap was purified by extracting the dried material with ether for 20 minutes using a Soxlet extractor. The less soluble fraction, the purified material, was then added to a polymerization. The material extracted by solvent was a mixture of pure material and some impurity, and was also added to a polymerization.

A comparison of the polymerization rates for pure and impure Sipex TDS is shown in Fig. 4. With the purified material, the rate of the polymerization is not retarded, while the extract shows a very low rate of polymerization. Thus, there was an impurity in the Sipex TDS which retarded polymerization. However, if the soaps are pure, the rate of polymerization is independent of the soap type at 0·5% concentration.

Since no dependence of polymerization rate on soap type was found, a detailed study of the three surfactants was no longer needed. Polystep B-19 was chosen as the emulsifier for this kinetic study because no report concerning vinyl acetate emulsion polymerization using an alkyl polyethoxysulfate soap had been found in the literature. For comparison and as a control, the dependencies of the polymerization rate on two parameters (initiator concentration and soap concentration) was also studied using sodium lauryl sulfate.

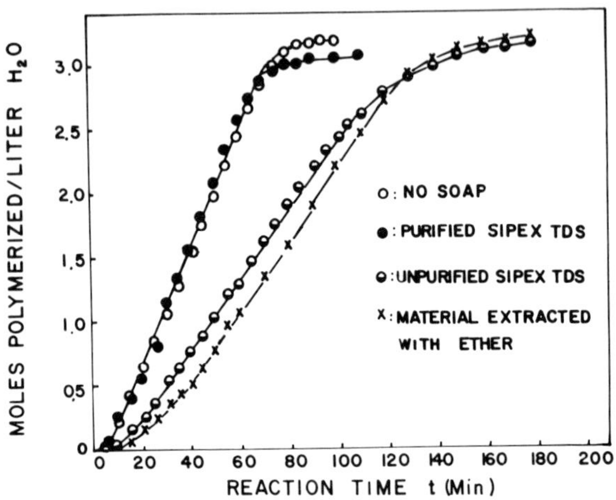

Fig. 4. The effect of impurity of Sipex TDS on the polymerization rate. (Soap concentration = 5·05 g/liter H_2O, $V_{org}/V_{aq} = 0{\cdot}33$, $(I_0) = 5 \times 10^{-4}$ M, $\mu = 3 \times 10^{-3}$.)

Dependence of rate on soap concentration

Experiments were conducted varying the emulsifier concentration from 0% to 1%, while the other parameters were kept constant. Both sodium lauryl polyether sulfate (Polystep B-19) and sodium lauryl sulfate (Stepanol WA-100, SLS) were used. It can be seen from Table V that there is little if any dependence of rate on either emulsifier.

TABLE V

Dependence of polymerization rate on soap concentration in seeded polymerizations†

Soap concentration (%)	0	0·1	0·25	0·50	1·0
100 Rp mole/liter min (Stepanol WA-100, SLS)	4·70	4·48	4·40	4·44	4·28
100 Rp mole/liter min (Polystep B-19)	4·70	4·62	4·40	4·29	4·41

† Initiator concentration 5×10^{-4} M, ionic strength 3×10^{-3}, particle concentration $7{\cdot}27 \times 10^{17}$/liter H_2O, $V_{org}/V_{aq} = 0{\cdot}33$.

Initiator dependence

The effect of initiator concentration on the rate of polymerization was determined over a wide range of concentrations. The initial persulfate concentration was varied from 1×10^{-4} M to 10×10^{-4} M and the particle concentration was varied from $7{\cdot}27 \times 10^{17}$/liter H_2O to $1{\cdot}04 \times 10^{17}$/liter H_2O. 0·25% of Polystep B-19 was added in the runs with higher particle concentration (7·27 and $3{\cdot}31 \times 10^{14}$/cc H_2O) and 0·1% in lower particle concentration (1·67 and $1{\cdot}04 \times 10^{14}$/cc H_2O) runs to keep the particles from coagulating. Potassium sulfate was added as needed to keep the ionic strength constant in each run at 3×10^{-3}.

Plots of monomer polymerized versus time as a function of initiator concentration are shown in Fig. 5. The rates of the polymerization were measured from the slope of the linear portion of the curves. A plot of log rate versus log initiator concentration for several particle concentrations is shown in Fig. 6. There is some scatter of the points; however, a slope of 0·60 is strongly indicated. Since the initiator concentration changes during the polymerization,[20] the average concentration between 30 and 90% conversion, the linear part of the polymerization, was used to correlate the results. This was measured separately.

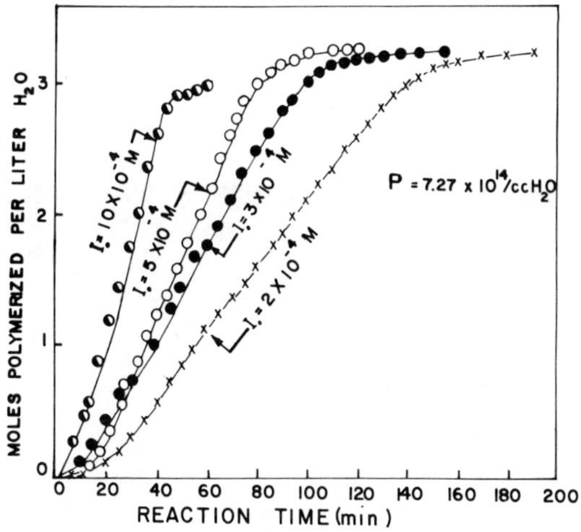

Fig. 5. The effect of initiator concentration on the polymerization rate ($V_{org}/V_{aq} = 0{\cdot}33$, $\mu = 3 \times 10^{-3}$).

The dependence of polymerization rate on initiator concentration when sodium lauryl sulfate (Stepanol WA-100) was used as soap was also studied. The experimental conditions were the same as before, but only one particle concentration was used. A log-log plot of the polymerization rate versus persulfate concentration is shown in Fig. 7; a slope of $0{\cdot}62 \pm 0{\cdot}02$ was found.

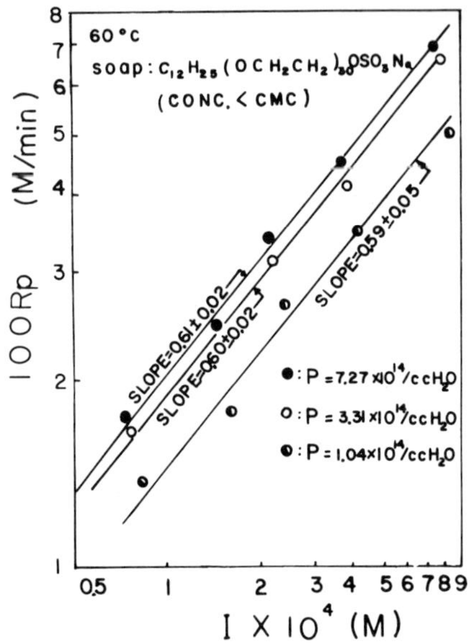

Fig. 6. Dependence of polymerization rate on initiator concentration at several particle concentrations (V_{org}/V_{aq} = 0·33, $\mu = 3 \times 10^{-3}$; CMC = critical micelle concentration).

Again, the average initiator concentration from 30% to 90% conversion was used. Thus the dependence of polymerization rate on persulfate concentration was 0·6 power for two types of soap.

Dependence of rate on volume of organic phase per unit volume of aqueous phase, V_{org}/V_{aq}

The relative volume of organic phase was varied from 0·103 to 0·46 with the other parameters kept constant. A log–log plot of the polymerization rate versus volume of organic phase is shown in Fig. 8. The rate is proportional to the volume of organic phase to 0·39 ± 0·08 power.

Variation of rate with ionic strength, μ

Potassium sulfate was added to change the ionic strength of the polymerization system while other parameters were kept invariant. The ionic strength was changed twentyfold from 3×10^{-3} to 60×10^{-3}. From Table VI, it can be seen that there is little if any dependence of rate on ionic strength.

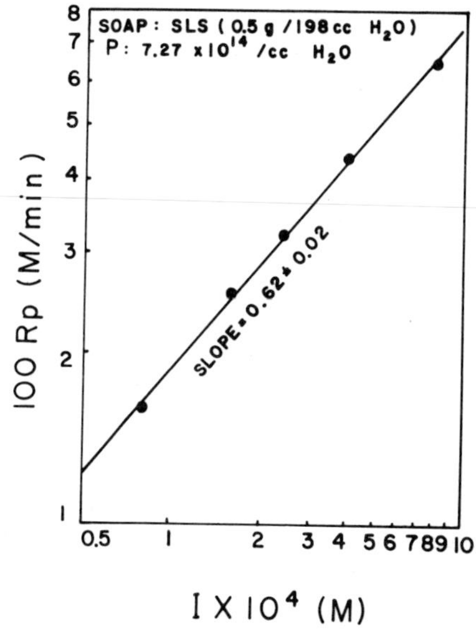

Fig. 7. The dependence of the polymerization rate on persulfate concentration when sodium lauryl sulfate was used as emulsifier ($(I_0) \approx 1$ to 10×10^{-4} M, $V_{org}/V_{aq} = 0{\cdot}33$, $\mu = 3 \times 10^{-3}$).

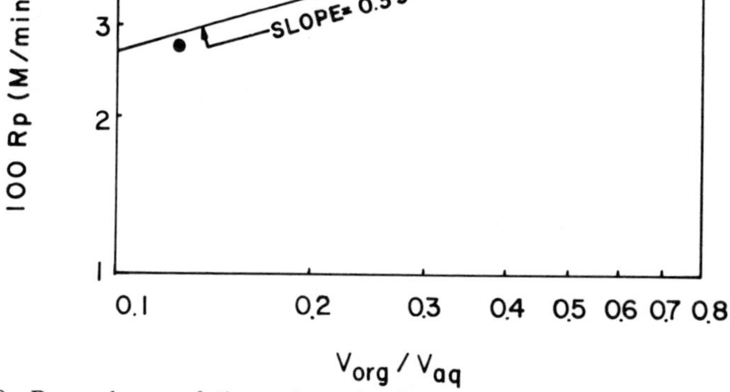

Fig. 8. Dependence of the polymerization rate on volume of organic phase per unit volume of aqueous phase (V_{org}/V_{aq}).

TABLE VI
Dependence of rate on ionic strength†

Run No.	17	34	35	36	37
Ionic strength	3×10^{-3}	6×10^{-3}	15×10^{-3}	30×10^{-3}	60×10^{-3}
100 Rp (mole/liter min)	4·48	4·50	4·48	4·24	4·20

† $(I_0) = 5 \times 10^{-4}$ M, soap (Polystep B-19) 0·5%, $(P) = 7{\cdot}27 \times 10^{17}$/liter H_2O.

Dependence of rate on particle concentration

This was studied by polymerizing at a given initiator concentration while varying the seed concentration. The initial seed particle concentration was calculated as $2{\cdot}99 \times 10^{18}$ particles/liter H_2O by eqn. (1) after the average particle diameter had been measured using quasi-elastic laser light scattering.[18] This was diluted with water to give initial particle concentrations of 6·72, 3·36, 1·68 and $0{\cdot}84 \times 10^{17}$/liter H_2O. Enough monomer was then added to give 25 wt% of polymer plus monomer. Small amounts of soap were added, as described earlier, to stabilize the particles. A determination of the particle concentrations before and after polymerization was necessary to see whether the particle number changed or not. Table VII lists the particle concentrations before and after polymerization for different runs.

TABLE VII
Particle concentration before and after polymerization

Run no.	*Before polymerization. Initial particle conc. (P_i) ($\times 10^{17}$/liter H_2O)*	*After polymerization Final particle conc. (P_f) ($\times 10^{17}$/liter H_2O)*	*Number average diameter D_n (μm)*
Seed-2	29·9	—	0·045
15	6·72	7·51	0·089
17	6·72	7·51	0·089
18	6·72	6·80	0·092
19	3·36	3·14	0·119
21	3·36	3·57	0·114
22	3·36	3·22	0·118
24	1·68	1·67	0·147
30	0·84	1·08	0·170
31	0·84	0·99	0·175

The number average diameters were measured using quasi-elastic laser light scattering[18] and the particle numbers were calculated based on eqn. (1) with known w (0·17 g/cc H_2O for seed-2 and 0·33 g/cc H_2O for the rest in Table

VII) and ρ (1·19 for poly(vinyl acetate)). The final particle concentrations for a given dilution were averaged and compared with the calculated particle concentrations before polymerization. For runs 15, 17 and 18 the initial number was $6{\cdot}72\times10^{17}$/liter H_2O while the average at the end was $7{\cdot}27\times10^{17}$/liter H_2O; for runs 19, 21 and 22 we found $3{\cdot}36\times10^{17}$/liter H_2O versus $3{\cdot}31\times10^{17}$/liter H_2O; for run 24 it is $1{\cdot}68\times10^{17}$/liter H_2O versus $1{\cdot}67\times10^{17}$/liter H_2O; for runs 30 and 31 it is $0{\cdot}84\times10^{17}$/liter H_2O versus $1{\cdot}04\times10^{17}$/liter H_2O. Since the experimental error in particle diameter measurement in quasi-elastic laser light scattering is about 3 to 5%,[18] a 10 to 15% difference in particle number is within the bounds of the technique. It can be seen that the particle numbers remain unchanged within experimental error before and after polymerization for all the runs except those at very low particle concentrations, runs 30 and 31.

This can be rationalized because at very low particle concentrations, the added soap (0·1%) is above its critical micelle concentration, 0·06%, and may be able to stabilize new particles. At higher particle concentrations, the soap is probably adsorbed on the particles. Even so, the particle number increases only 24% for runs 30 and 31. These final average particle concentrations were used for the kinetic study. A plot of log rate versus log particle concentration after polymerization is shown in Fig. 9. Within experimental error, the slopes of the three sets of experiments agree. The average dependence of rate on particle concentration is 0·12 ± 0·03.

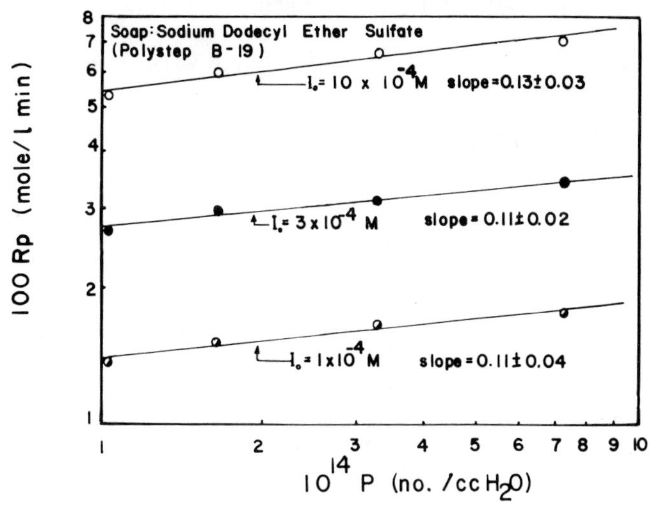

Fig. 9. Dependence of the polymerization rate on particle concentration ($V_{org}/V_{aq}=0{\cdot}33$, $\mu=3\times10^{-3}$).

Table VIII summarizes the experimental order of dependence of the polymerization rate on the experimental parameters. The polymerization rate in this work is independent of ionic strength and depends on initiator concentration to the 0·6 power. This is more in line with the results of Nomura *et al.*[2] and Friis and Nyhagen[1] but different from the results of Patsiga and co-workers.[3,4] There, the rate of polymerization was dependent on initiator concentration to the 0·8 power and was thought to depend on ionic strength of the aqueous solution to the 0·3 power. (The high power for initiator concentration dependence is probably because impure soap was used which retarded the polymerization and gave a higher power dependency of rate on initiator concentration. This will be discussed in a separate paper.) Since the Litt–Patsiga–Stannett model was inconsistent with the lack of dependence of rate on soap type and ionic strength, a new model had to be developed to account for what was happening.

TABLE VIII
Summary of experimental results: order of dependence of polymerization rate on experimental variables

$$Rp = k \prod_{i=1}^{6} Y_i^{n_i}$$

$(Y = (I), (P), V_{org}, \mu$, soap conc., type of soap)

Variable	$n_{experimental}$
(I)	$0{\cdot}60 \pm 0{\cdot}03$
(P)	$0{\cdot}12 \pm 0{\cdot}03$
V_{org}/V_{aq}	$0{\cdot}39 \pm 0{\cdot}08$
μ	0
Soap conc.	0
Type of soap	0

DISCUSSION

Proposed model

Based on the experimental results summarized in Table VIII, the general chemistry of active radicals and the physical chemistry of emulsions, a kinetic model is proposed. This presents the reaction steps at a molecular level for vinyl acetate emulsion polymerization and derives the kinetic equations for the polymerization. It does not include particle formation. In many steps it is close to the models proposed by Friis and Nyhagen[1] and Nomura *et al.*[2] However, it differs in important ways, which will become obvious.

1. $S_2O_8^=$ decomposes into two $SO_4^-\cdot$ groups which enter particles. If the particle is dead, the $SO_4^-\cdot$ initiates polymerization, $\rightarrow P\cdot (P\cdot \equiv P\bar{n})$. If the particle contains a radical, it terminates polymerization.
2. The growing radical in a particle, $P\cdot$, adds monomer in a normal propagation step.
3. The radical, $P\cdot$, chain transfers to monomer, $\rightarrow M_1\cdot$. The subscript 1 shows that the monomer radical is in its original particle.

We propose that chain transfer is mainly on the vinyl group. Analysis of chain transfer constants from literature[13–16] shows that most people believe that the major chain transfer site is on the acetyl hydrogens of vinyl acetate. But we now have shown that the major and kinetically significant chain transfer site is on the vinyl hydrogens of vinyl acetate.[21] Also, there is evidence in the literature[22–25] to show that the vinyl radical is a stable radical. We will discuss this in more detail later.

The monomer radical, perhaps $CH_2{=}C\cdot{-}OAc$, reinitiates relatively slowly compared to the normal propagation rate, and thus the radical can escape easily from the particle to become an aqueous radical, $M\cdot_a$. It may also reinitiate polymerization in the initial particle. (Some chain transfer on the acetyl group occurs, but this should be kinetically insignificant since that radical reinitiates promptly before it can diffuse out of the particle. If this were the only chain transfer step, vinyl acetate would behave like styrene.)

4. The aqueous radical, $M\cdot_a$, can do several things:
   (a) It can be swept up by a dead particle, generating $M\cdot_p$.
   (b) It can react with initiator to generate a $SO_4^-\cdot$ radical ion which will then continue as in step 1.
   (c) Aqueous monomer radical can be swept up into a particle containing a radical, $\rightarrow (M\cdot_p P\cdot)$, which probably terminates. However, because the time it stays in a particle is relatively short, the radical may diffuse out of such a particle without terminating; this is included in the model.
5. The rate of diffusion of a small molecule out of a particle, if we assume no interfacial barrier, is calculated in Appendix A to be $4D/ad^2$. (D is the diffusion constant of the radical in the monomer/polymer phase, a is the partition coefficient and d is the diameter of the particle.)
6. $M\cdot_p$ will either add monomer, $\rightarrow P\cdot$, or diffuse again into the aqueous phase, $\rightarrow M\cdot_a$.

Nature of chain transfer to the monomer

As was mentioned, most researchers believe that the acetyl hydrogens are the main chain transfer site.[13,14] Although Wheeler *et al.*[15] and Howard[16] found a chain transfer ratio of vinyl to acetyl hydrogens of 0·33 to 0·5, this still implies that most chain transfer is on the acetyl hydrogens. Litt *et al.*[4] had proposed that the acetyl chain transferred monomer radicals would cyclize to form a butyrolactonyl radical which reinitiates as fast or faster than the growing radical. A calculation of the probability of escape based on the formula in Friis' thesis[1b] and Appendix A here shows that the probability of escape of butyrolactonyl radical is very small; it reinitiates rapidly compared to diffusion out of particles. Thus chain transfer on the acetyl hydrogens gives a species which cannot affect the kinetics.

The other possibility is chain transfer on vinyl hydrogens. There is evidence to show that the vinyl radical is more stable than the corresponding alkyl radical.[22–25] Ryska *et al.*[22] have found that addition of propionaldehyde in the homogeneous polymerization of vinyl chloride in solution increases the rate of the polymerization. They interpret this as chain transfer of a stable radical ($CH_2{=}\dot{C}Cl$) on the aldehyde, generating an active radical which reinitiates rapidly. Breitenbach and co-workers[23,24] have also observed an increase of polymerization rate of vinyl chloride by addition of tetrabromomethane or dodecyl mercaptan to the polymerizing system. This was explained by postulating reactivation of the less reactive vinyl chloride radical.

Higashiura and Oiwa[25] have studied the free radical polymerization of styrene and methyl methacrylate in the presence of phenyl acetylene and five of its *p*-substituted derivatives. They found that phenyl acetylene retarded the polymerization. This was rationalized by postulating the radical, $\sim CH{=}\dot{C}{-}\phi$, produced by copolymerization is very stable and terminates by reaction with the growing radical. This implies that the vinyl radical is stable compared to the growing radical. Since chain transfer on the vinyl hydrogens of vinyl chloride and ethylene is of the same magnitude as the vinyl acetate chain transfer constant, transfer on the vinyl hydrogens of vinyl acetate is reasonable. (This will be discussed in more detail later.[21] There, it is shown that the main chain transfer site of vinyl acetate is the vinyl hydrogens (94%).)

The kinetic scheme used, based on the description above, is given below.

	Chemical reaction	*Description*	*Kinetic rate term*
1.	$I \rightarrow 2fP\cdot(1-2\bar{n})$	Initiation and Smith–Ewart termination	$2fk_1(I)(1-2\bar{n})$
2.	$P\cdot + M \rightarrow P\cdot$	Propagation	$k_2(M)(P\bar{n})$

3.	$P\cdot + M \rightarrow M_1$	Chain transfer to monomer	$k_3(M)(P\bar{n})$
4.	$M_1\cdot + M \rightarrow P\cdot$	Reinitiation in same particle	$k_4(M)(M_1\cdot)$
5.	$M_1\cdot \rightarrow M\cdot_a$	Diffusion (escape) of monomer radical from initial particle†	$(5D/ad^2)(M_1\cdot)$
5a.	$M\cdot_p \rightarrow M\cdot_a$	Escape of monomer radical from particle†	$(4D/ad^2)(M\cdot_p)$
6.	$M\cdot_a + P(1-\bar{n}) \rightarrow M\cdot_p$	Sweep up of aqueous radical into dead particle	$2\pi D_a d(P)(1-\bar{n})(M\cdot_a)$
6a.	$M\cdot_a + P\cdot \rightarrow (M\cdot_p P\cdot)$	Sweep up of aqueous radical into occupied particle	$2\pi D_a d(P\bar{n})(M\cdot_a)$
4a.	$M\cdot_p + M \rightarrow P\cdot$	Reinitiation in secondary particle	$k_4(M)(M\cdot_p)$
4b.	$(M\cdot_p P\cdot) + M \rightarrow$ Termination	Reinitiation of growth and termination of the two radicals in the particle	$k_4(M)(M\cdot_p P\cdot)$
5b.	$(M\cdot_p P\cdot) \rightarrow M\cdot_a + P\cdot$	Escape of monomer radical from occupied particle†	$(4D/ad^2)(M\cdot_p P\cdot)$
7.	$M\cdot_a + I \rightarrow P\cdot(1-2\bar{n})$	Induced decomposition of initiator, and sweep up of $SO_4^-\cdot$ radical in particle with initiation	$k_7(I)(M\cdot_a)(1-2\bar{n})$
8.	$(M\cdot_p P\cdot) \rightarrow$ Termination	Termination of $M\cdot_p$ with $P\cdot$	$\dfrac{6k_{80}D}{N\pi d^3 D_0}(M\cdot_p P\cdot)$

† The equation given for steps 5, 5a and 5b is derived in Appendix A.

Steady state equations

$R_i \equiv$ Rate of step i

$$\frac{d(M_1\cdot)}{dt} = R_3 - R_4 - R_5 = 0 \tag{2}$$

$$\frac{d(M\cdot_p)}{dt} = R_6 - R_{4a} - R_{5a} = 0 \tag{3}$$

$$\frac{d(M\cdot_p P\cdot)}{dt} = R_{6a} - R_{5b} - R_8 = 0 \tag{4}$$

$$\frac{d(M\cdot_a)}{dt} = R_5 + R_{5a} + R_{5b} - R_6 - R_{6a} - R_7 = 0 \tag{5}$$

$$\frac{\mathrm{d}(P\cdot)}{\mathrm{d}t} = R_1 + R_4 + R_{4a} + R_{5b} + R_7 - R_3 - R_{6a} = 0 \tag{6}$$

Step 4b is omitted from eqn. (4) since the chance of the reinitiation of monomer radical in a particle that already contains a radical is much less than termination.

The rate of initiation is equal to the rate of termination under steady state conditions. Thus:

$$R_1 = R_8 \tag{7}$$

Some further definitions are needed. The probability of $(M\cdot_{\mathrm{p}})$ diffusing out of a particle as compared to reinitiating growth is P_{e}, the probability of escape.

$$P_{\mathrm{e}} = R_{5a}/(R_{4a} + R_{5a}) = 1/(1 + R_{4a}/R_{5a}) \tag{8}$$

The probability of $(M\cdot_{\mathrm{p}}P\cdot)$ terminating rather than $(M\cdot)$ escaping is P_{t}, the probability of termination.

$$P_{\mathrm{t}} = R_8/(R_8 + R_{5b}) = 1/(1 + R_{5b}/R_8) \tag{9}$$

The equations for the concentration of the various radicals in terms of $(M\cdot_{\mathrm{a}})$, $(P\cdot)$ or $\bar{n}$ are:

$$(M_1\cdot) = (k_3/k_4)\,(1 - P_{\mathrm{e}})\,(P\cdot) \tag{10}$$

$$(M\cdot_{\mathrm{p}}) = \frac{2\pi D_{\mathrm{a}} d(P)(1-\bar{n})}{4D/ad^2}\, P_{\mathrm{e}}(M\cdot_{\mathrm{a}}) \tag{11}$$

$$(M\cdot_{\mathrm{p}}P\cdot) = \frac{2\pi D_{\mathrm{a}} d}{4D/ad^2}\,(1 - P_{\mathrm{t}})(P\cdot)(M\cdot_{\mathrm{a}}) \tag{12}$$

$$(M\cdot_{\mathrm{a}}) = \frac{(4D/ad^2)(k_3/k_4)(1-P_{\mathrm{e}})(P\cdot)}{2\pi D_{\mathrm{a}} d\,[(P)(1-\bar{n}) + (P\bar{n}) - (P)(1-\bar{n})P_{\mathrm{e}} - (1-P_{\mathrm{t}})(P\cdot) + k_7(I)(1-2\bar{n})/2\pi D_{\mathrm{a}} d]} \tag{13}$$

$$(M\cdot_{\mathrm{a}}) \approx \frac{(4D/ad^2)(k_3/k_4)(1-P_{\mathrm{e}})(P\cdot)}{2\pi D_{\mathrm{a}} d\,[(P)(1-P_{\mathrm{e}}) + k_7(I)(1-2\bar{n})/2\pi D_{\mathrm{a}} d]} \tag{14}$$

From eqn. (8),

$$P_e = \frac{1}{1 + R_{4a}/R_{5a}} = \frac{(4D/ad^2)}{(4D/ad^2) + k_4(M)} \tag{15}$$

Thus,

$$1 - P_e = \frac{R_{4a}/R_{5a}}{1 + R_{4a}/R_{5a}} = \frac{k_4(M)}{4D/ad^2 + k_4(M)} \tag{16}$$

Combining (15) and (16):

$$\frac{P_e}{1 - P_e} = \frac{4D/ad^2}{k_4(M)} \tag{17}$$

Substituting (17) in (14) yields

$$(M\cdot_a) = \frac{k_3 P_e (M)(P\cdot)}{2\pi D_a d(P)\,[(1 - P_e) + k_7(I)(1 - 2\bar{n})/2\pi D_a d(P)]} \tag{18}$$

From (9)

$$P_t = \frac{1}{1 + R_{5b}/R_8} = \frac{\dfrac{6k_{80}D}{N\pi d^3 D_0}}{\dfrac{6k_{80}D}{N\pi d^3 D_0} + \dfrac{4D}{ad^2}} \tag{19}$$

$$1 - P_t = \frac{4D/ad^2}{\dfrac{6k_{80}D}{N\pi d^3 D_0} + \dfrac{4D}{ad^2}} = \frac{4N\pi d D_0}{6ak_{80} + 4N\pi d D_0} \tag{20}$$

From (7), $R_1 = R_8$

$$2fk_1(I)(1 - 2\bar{n}) = \frac{6k_{80}D}{N\pi d^3 D_0}(M\cdot_p P\cdot) \tag{21}$$

Substituting (12), (18), and (20) into (21), yields:

$$(P\cdot)^2 + (P\cdot)\frac{4fk_1(I)\left[(1 - P_e) + \dfrac{k_7(I)}{2\pi D_a d(P)}\right]}{k_3 P_t P_e (M)} = \frac{2fk_1(I)(P)\left[(1 - P_e) + \dfrac{k_7(I)}{2\pi D_a d(P)}\right]}{k_3 P_t P_e (M)} \tag{22}$$

We can now solve approximately for $(P\cdot)$ if we assume that the Smith–Ewart radical termination step is small compared to step 8. Later calculations show that at these particle concentrations, it becomes important only above 90% conversion.

$$(P\cdot)\approx\left(\frac{2fk_1(I)(P)\left[(1-P_e)+\dfrac{k_7(I)}{2\pi D_a d(P)}\right]}{k_3P_tP_e(M)}\right)^{1/2}-\frac{2fk_1(I)\left[(1-P_e)+\dfrac{k_7(I)}{2\pi D_a d(P)}\right]}{k_3P_tP_e(M)}$$

$$=\left(\frac{2fk_1(I)(P)\left[(1-P_e)+\dfrac{k_7(I)}{2\pi D_a d(P)}\right]}{k_3P_tP_e(M)}\right)^{1/2}\left(1-\frac{fk_1(I)\left[(1-P_e)+\dfrac{k_7(I)}{2\pi D_a d(P)}\right]^{1/2}}{(fk_1k_3P_eP_t(M)(I)(P))^{1/2}}\right) \tag{23}$$

where

$$\frac{fk_1(I)\left[(1-P_e)+\dfrac{k_7(I)}{2\pi D_a d(P)}\right]^{1/2}}{(fk_1k_3P_eP_t(M)(I)(P))^{1/2}}$$

is an approximation for the fraction of radicals terminated by the Smith–Ewart mechanism when it is less than 50% of the total. Substituting (17) into (23), we get

$$(P\cdot)\simeq(2fk_1(I)(P))^{1/2}(k_4/k_3)^{1/2}(ad^2/4D)^{1/2}(1/P_t)^{1/2}$$
$$\times\left(1+\frac{4Dk_7(I)}{2\pi k_4D_ad^3(P)P_e(M)}\right)^{1/2}\left(1-\frac{fk_1(I)\left[(1-P_e)+\dfrac{k_7(I)}{2\pi D_a d(P)}\right]^{1/2}}{(fk_1k_3P_eP_t(M)(I)(P))^{1/2}}\right) \tag{24}$$

The rate of polymerization is

$$Rp=k_2(M)(P\cdot) \tag{25}$$

The ratio of organic volume to aqueous volume is

$$V_{org}/V_{aq}=(P)VN=\frac{\pi d^3(PN)}{6} \tag{26}$$

where N is Avogadro's number, V is the volume of a particle of diameter d; and d is in units of decimeters. Substituting (19), (24) and (26) into (25), the rate of polymerization is then written as eqn. (27).

$$Rp = k_2(M)(P\cdot) = (6/\pi)^{1/3}(1/N^{1/3})\, k_2(fk_1k_4/2k_3)^{1/2}(1 + 2\pi D_0\, \mathrm{d}N/3ak_{80})^{1/2}$$

$$\times\, a^{1/2}((M)/D^{1/2})\,(V_{\mathrm{org}}/V_{\mathrm{aq}})^{1/3} P^{1/6}(I)^{1/2}\left[1 + \frac{4k_7D(I)}{2\pi k_4 D_{\mathrm{a}} d^3(P)\, P_{\mathrm{e}}(M)}\right]^{1/2}$$

$$\times\left(1 - \frac{fk_1(I)\left[(1-P_{\mathrm{e}}) + \dfrac{k_7(I)}{2\pi D_{\mathrm{a}} d(P)}\right]^{1/2}}{[fk_1k_3P_{\mathrm{e}}P_{\mathrm{t}}(M)(I)(P)]^{1/2}}\right) \qquad (27)$$

Equation (27) predicts that rate of polymerization is proportional to $(V_{\mathrm{org}}/V_{\mathrm{aq}})$ to the $0{\cdot}33^{+}$ power, particle concentration to the $0{\cdot}167^{-}$ power, initiator concentration to the $0{\cdot}5^{+}$ power, monomer to the 1^{+} power and diffusion constant of monomer radical in the particle to the $-\frac{1}{2}$ power, and is independent of soap concentration and ionic strength. In order to get a final workable equation, D and a must be expressed as a function of conversion.

Diffusion coefficient of monomer radical in the particle as a function of monomer concentration

Goffloo and Kosfeld measured[26,27] the diffusion coefficient of benzene in polystyrene as a function of volume fraction of benzene from 0·2 to 0·7 at different temperatures using NMR spin echo measurements. We have taken their data and plotted the square root of the self diffusion coefficient of benzene versus the concentration of benzene in polystyrene, Fig. 10. As can be seen, a straight line relationship was found for the temperature range of 60–100°C. $D^{1/2}$ approaches zero at finite benzene concentration, presumably because the system is nearing Tg. When the temperature is about 110°C, 15-20°C above the Tg of polystyrene, D should be zero when (M) is zero. This means $D^{1/2}$ would be exactly proportional to the concentration of benzene in the polystyrene at this temperature.

Fujita[28] measured the diffusion coefficient of ethyl acetate in polymethyl acrylate with volume fractions of ethyl acetate from 0·02 to 0·11, by sorption–desorption measurements. His data were taken and replotted in Fig. 11 as the square root of the diffusion coefficient of ethyl acetate versus its concentration. Again the data fall on straight lines, with intercepts that bracket zero at 20–30°C above Tg.

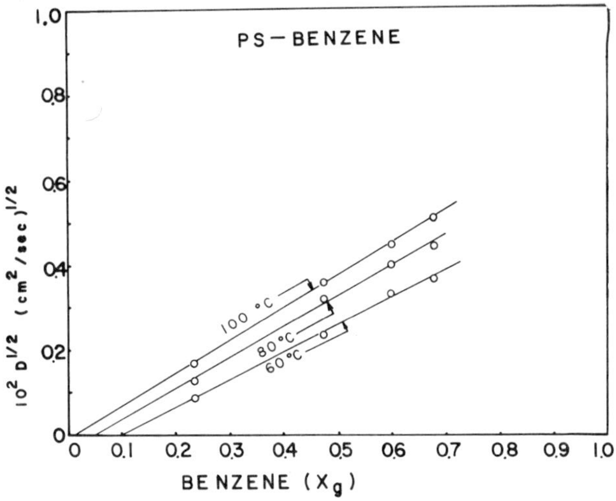

Fig. 10. The plot of the square root of the diffusion coefficient of benzene versus the concentration of benzene in polystyrene (Goffloo and Kosfeld[26, 27]) X_g is the weight fraction of benzene.

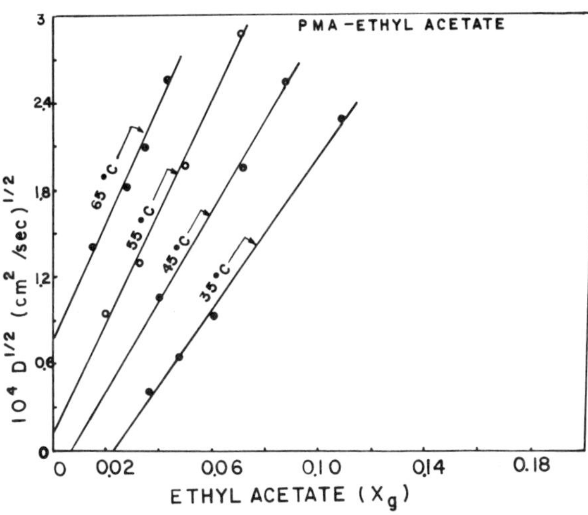

Fig. 11. The plot of the square root of the diffusion coefficient of ethyl acetate versus the concentration of ethyl acetate in polymethyl acrylate (Fujita[28]), X_g is the weight fraction of ethyl acetate.

In Friis' thesis,[1b] he obtained the diffusion coefficient of monomer radical in the particle at 50°C as a function of polymer conversion based on the computation of polymer conversion data versus the probability of escape, which contains the diffusion coefficient. Again a plot of $D^{1/2}$ versus (M) is linear. A log–log plot of diffusion constant versus monomer concentration is shown in Fig. 12; a slope of 2·1 is found. At 60°C, the slope would be about 2·0.

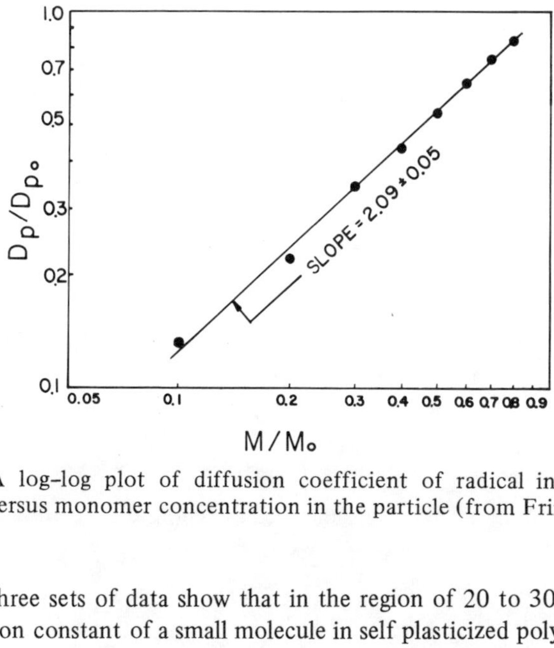

Fig. 12. A log–log plot of diffusion coefficient of radical in the particle versus monomer concentration in the particle (from Friis[1]).

Thus, three sets of data show that in the region of 20 to 30°C above *Tg*, the diffusion constant of a small molecule in self plasticized polymer is about proportional to the square of the plasticizer concentration. As no contradictory data were found, the approach is interesting. It gives a very simple way of writing the diffusion constant dependence, $D^{1/2} = K(M - M_0)$, where M is the volume fraction of the small molecule, in the region of 40°C below to 60°C above the *Tg*. M_0 is zero at about 20–30°C above *Tg*, positive at lower temperatures and negative at higher temperatures.

Definition of partition function

If we consider that there is no barrier to a monomer molecule or radical diffusing from water into a particle, then the positive change in interaction energy on diffusion from a particle into water will increase the time a radical is held within the particle by the ratio of the solubilities of the monomer in the two phases. This is the minimum increase in time, since we are assuming no interfacial barrier. As the chain transferred vinyl radical is very similar to monomer, we use the monomer partition coefficient,

$$a = \frac{(M)}{(M_a)} \tag{28}$$

where (M_a) is the monomer concentration in aqueous phase.

Over most of the polymerization range (M_a) was found[4, 7, 27] to be proportional to $(M)^{1/2}$. Thus:

$$(M_a) = (M_{a_0}) \left[\frac{(M)}{(M_0)}\right]^{1/2} \tag{29}$$

where (M_{a_0}) and (M_0) are the monomer concentrations in the aqueous and organic phases, respectively, before the disappearance of the separate monomer phase. At 60°C, (M_{a_0}) was found[4] to be 0·30 mole/liter, while (M_0) has been found in different studies to range from 6·3 to 7·6 mole/liter in polymer particles.[3, 30] Here we take the average for (M_0) as 7 mole/liter. Thus $a = 0{\cdot}113/(M)^{1/2}$.

The probability of termination, P_t, involves escape from the particle and includes the partition coefficient. Thus, from eqns. (19) and (29)

$$P_t^{-1/2} = \left(1 + \frac{0{\cdot}237 D_0 dN}{k_{80} \cdot (M)^{1/2}}\right)^{1/2} \tag{30}$$

After combining all the constants and including the partition coefficient of the monomer radical explicitly, we obtain

$$\begin{aligned} Rp = {} & 3{\cdot}09 \times 10^{-8} \left(\frac{k_2 (f k_1 k_4)^{1/2}}{k_3^{1/2}}\right) \left(1 + \frac{1{\cdot}42 \times 10^{23} D_0 d}{k_{80} \cdot (M)^{1/2}}\right)^{1/2} \\ & ((M)^{5/4}/D^{1/2}) (V_{org}/V_{aq})^{1/3} (P)^{1/6} (I)^{1/2} \left(1 + \frac{0{\cdot}072 D k_7 (I)}{k_4 D_a d^3 (P) P_e (M)^{3/2}}\right)^{1/2} \\ & \times \left(1 - \frac{f k_1 (I) \left[(1 - P_e) + \dfrac{k_7 (I)}{2\pi D_a d (P)}\right]^{1/2}}{(f k_1 k_3 P_e P_t (M)(I)(P))^{1/2}}\right) \end{aligned} \tag{31}$$

We have presented evidence that $D^{1/2}$ should be proportional to (M) at our experimental conditions. Thus:

$$D^{1/2} = D_0^{1/2} \frac{(M)}{(M_0)} \tag{32}$$

The standard termination step of the Smith–Ewart model is very much smaller than step 8 of this model. It will be shown later to be between 0·007 and 0·12 at 90% conversion for all but one case, and thus will be neglected as it is within experimental error.

Substituting (32) into (31) and omitting the Smith–Ewart termination term, the final equation for the rate of polymerization is

$$Rp = 3{\cdot}09 \times 10^{-8}\,\frac{k_2(fk_1k_4)^{1/2}}{k_3^{1/2}}\,((M_0)/D_0^{1/2})\left(1+\frac{1{\cdot}42\times 10^{23}D_0 d}{k_{80}\cdot(M)^{1/2}}\right)^{1/2}$$

$$\cdot\,(M)^{1/4}\cdot(V_{org}/V_{aq})^{1/3}\,(P)^{1/6}\,(I)^{1/2}\left(1+\frac{0{\cdot}072k_7D_0(M)^{1/2}(I)}{k_4D_ad^3(P)P_e(M_0)^2}\right)^{1/2} \quad (33)$$

For simplicity, eqn. (33) can be written as

$$Rp = C_1(1 + C_2d/(M)^{1/2})^{1/2}(M)^{1/4}(V_{org}/V_{aq})^{1/3}(P)^{1/6}(I)^{1/2}(1 + C_3(I)(M)^{1/2})^{1/2} \quad (34)$$

where

$$C_1 = 3{\cdot}09 \times 10^{-8}\left(\frac{k_2(fk_1k_4)^{1/2}}{k_3^{1/2}}\right)((M_0)/D_0^{1/2}) \quad (35)$$

$$C_2 = \frac{1{\cdot}42 \times 10^{23}D_0}{k_{80}} \quad (36)$$

$$C_3 = \frac{0{\cdot}072k_7D_0}{k_4D_ad^3(P)\,P_e(M_0)^2} \quad (37)$$

Equation (33) can be integrated as shown in Appendix B. The C_1, C_2 and C_3 values which give the best fits between the calculated curves and the experimental data are shown in Table IX.

TABLE IX
Values for C_1, C_2 and C_3 which give the best fit between the theory and experimental data

C_1 (s liter$^{1/3}$)	0·236
C_2 (mole/liter$^{1/3}$)	$1{\cdot}33 \times 10^6$
C_3 (liter/mole)[a]	407

[a] At $d = 0{\cdot}117\ \mu$m, $(P) = 3{\cdot}31 \times 10^{17}$/liter H_2O (average values).

Comparison of the experimental and theoretical orders of dependence of polymerization rate on various parameters

The expression for the rate of emulsion polymerization of vinyl acetate based on our kinetic model is given in eqn. (33). *Rp* is approximately independent of monomer concentration and an approximate straight line is found between 30 and 80% conversion when polymer conversion is plotted as a function of time. Since the parameters such as (*P*) appear several times, the dependence is not simple. The best way to find the dependence is to calculate rates by substituting various values for one parameter, keeping all others constant, using the best values of C_1, C_2 and C_3. A log–log plot of calculated rate versus the parameter varied gives the theoretical order. Table X summarizes the comparison between the theoretical predictions and experimental results. The fit is essentially perfect.

TABLE X
Comparison of the experimental and theoretical orders (n) of polymerization rate dependence on the experimental variables

Parameter	n_{exp}	n_{calc}
(I)	0·60 ± 0·03	0·60 ± 0·016
(P)	0·12 ± 0·03	0·094 ± 0·002
V_{org}/V_{aq}	0·39 ± 0·08	0·387 ± 0·002
μ	0	0
Soap concentration	0	0
Type of soap	0	0

Agreement of eqn. (33) with the polymerization curves

If eqn. (33) is meaningful, in addition to correlating the slopes of the linear portion of the curves it should predict the total course of polymerization under any given set of experimental conditions.

The course of polymerization can be divided into two parts. In the first part, a separate monomer phase is present. During this time, there is approximate equilibrium between $[M_a]$ and $[M]$ as they are replenished by fresh monomer. As the polymerization proceeds, the only variable is the particle size (the volume of organic phase); both the volume and surface area of the particle increase. (See details in Appendix B.) The integrated equation for this part is derived in Appendix B and is as follows.

$$\left(\frac{(M_0') - (M')}{b(M_0')}\right)^{2/3} - \left(\frac{V_{\mathrm{org_i}}}{V_{\mathrm{org_f}}}\right)^{2/3}$$
$$= \frac{2C_1(1 + C_2 d/(M_0)^{1/2})^{1/2}(M_0)^{1/4}(P)^{1/6}(I)^{1/2}(1 + C_3(I)(M_0)^{1/2})^{1/2}\, t}{3(M_0')\, b} \tag{38}$$

where $V_{\mathrm{org_i}}$ and $V_{\mathrm{org_f}}$ are the volumes of swollen seed and final organic phase respectively, (M_0') and (M') denote the moles of monomer per liter of H_2O at the beginning of polymerization and during polymerization, respectively, and b is the fractional conversion at which monomer phase disappears.

In the second part of the polymerization, the separate monomer phase has been used up. The particle volumes are essentially constant as is their surface area. Here the only variable is the monomer concentration. This portion of the equation has been derived in Appendix B also and is given in eqn. (39).

$$\left(0{\cdot}84 + \frac{C_2 d\left(\frac{V_{\mathrm{org}}}{V_{\mathrm{aq}}}\right)^{1/2}}{(M_0')^{1/2}}\right)^{1/2}\left(0{\cdot}28 - \frac{2}{3}\,\frac{C_2 d\left(\frac{V_{\mathrm{org}}}{V_{\mathrm{aq}}}\right)^{1/2}}{(M_0')^{1/2}}\right)$$
$$-\left(\left(\frac{(M')}{(M_0')}\right)^{1/2} + \frac{C_2 d\left(\frac{V_{\mathrm{org}}}{V_{\mathrm{aq}}}\right)^{1/2}}{(M_0')^{1/2}}\right)^{1/2}\left(\frac{1}{3}\left(\frac{(M')}{(M_0')}\right)^{1/2} - \frac{2}{3}\,\frac{C_2 d\left(\frac{V_{\mathrm{org}}}{V_{\mathrm{aq}}}\right)^{1/2}}{(M_0')^{1/2}}\right)$$
$$= \frac{C_1(V_{\mathrm{org}}/V_{\mathrm{aq}})^{1/12}(P)^{1/6}(I)^{1/2}(1 + C_3(I)(M)^{1/2})^{1/2}(t - t_{\mathrm{i}})}{4(M_0')^{3/4}} \tag{39}$$

where t_{i} is the time at which the separate monomer phase disappears.

Figures 13 and 14 show the fit between the experimental data and theoretically integrated curves where the initial initiator concentration was varied from 1×10^{-4} M to 10×10^{-4} M; all the other parameters are kept constant. As can be seen, the integrated curves follow the data over the whole conversion range for all different initiator concentrations.

Figure 15 shows the fit between the experimental data and the integrated curves for the extreme particle concentrations, $7{\cdot}27 \times 10^{17}$/liter H_2O and $1{\cdot}04 \times 10^{17}$/liter H_2O. The runs with other particle concentrations fit as well.

When $V_{\mathrm{org}}/V_{\mathrm{aq}}$ was varied, the fit between the experimental data and theoretical curves was also excellent as is shown in Fig. 16. Thus, the integrated theoretical curves fit very well with the experimental data for all the parameters over their whole range.

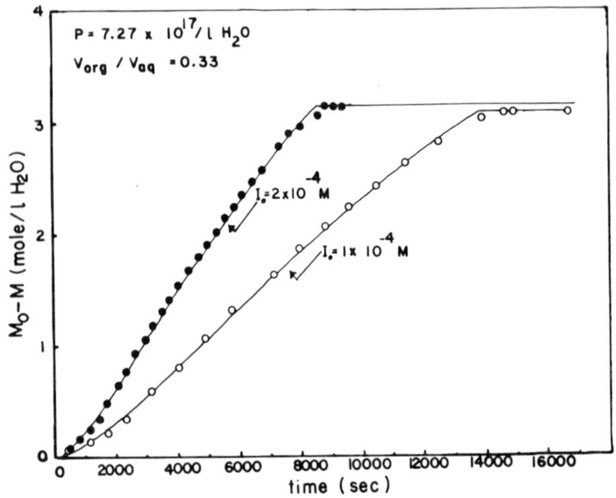

Fig. 13. Comparison of experimental conversions and integrated curves at low initiator concentrations, $(I_0) = 1 \times 10^{-4}$ M and 2×10^{-4} M.

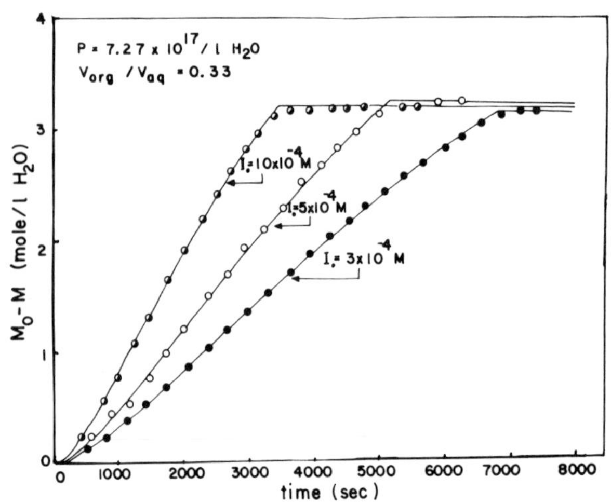

Fig. 14. Comparison of the experimental conversions and integrated curves. (I_0) varied from 3×10^{-4} M to 10×10^{-4} M.

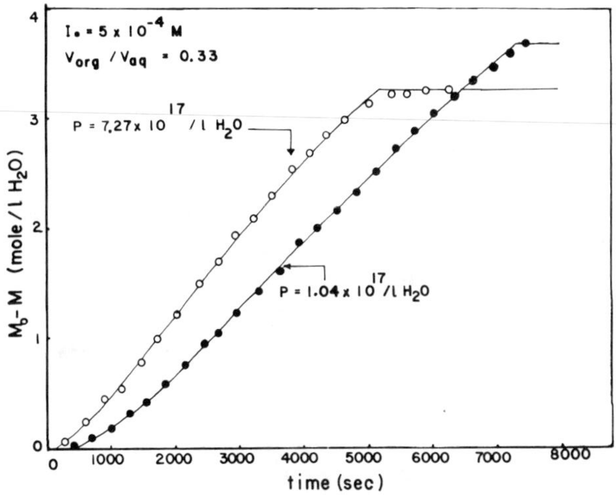

Fig. 15. The comparison of the experimental data with theoretical curves for two different particle concentrations.

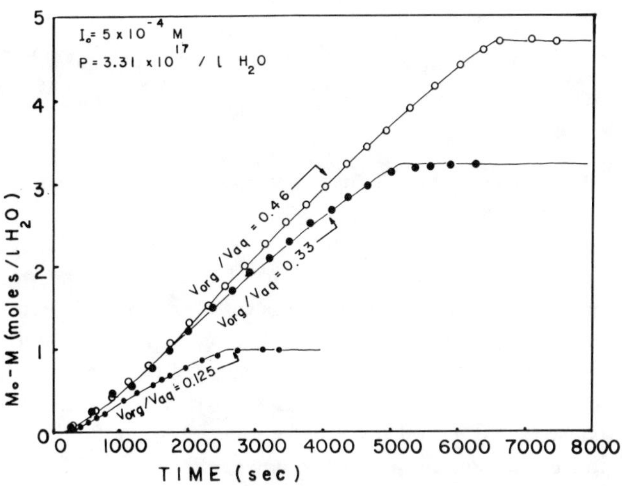

Fig. 16. The fit of the theoretical curves to the experimental data for different volumes of the organic phase.

Calculation of rate constants, k_4 and k_{80}

From Table IX, the best value for C_1 is 0·236 s liter$^{1/3}$. Once C_1 is known, we can calculate k_4, the monomer radical reinitiation rate constant, from eqn. (35).

$$C_1 = 3{\cdot}09 \times 10^{-8} \left(\frac{k_2(fk_1k_4)^{1/2}}{k_3^{1/2}}\right) ((M_0)/D_0^{1/2}) \tag{35}$$

k_1 is the unimolecular decomposition of persulfate at 60°C, $8{\cdot}3 \times 10^{-6}$/s.[31] The propagation rate constant at 60°C, k_2, in the literature ranges from 7×10^3 liter/mole s to 19×10^3 liter/mole s.[32a] We have taken a value of 10×10^3 liter/mole s for the calculations; the true k_2 is probably within 30% of this value. The chain transfer constant on monomer, k_3, is then 2·0 liter/mole s.[32b] (M_0) is the monomer concentration in the particle before the disappearance of separate monomer phase, 7 mole/liter H_2O. D_0 is taken from Friis' data[1] to be $1{\cdot}0 \times 10^{-8}$ dm^2/s. The efficiency in the unimolecular decomposition of potassium persulfate, f, is approximately 1. k_4 is then calculated to be 29 liter/mole s. (If k_2 is taken as 7×10^3 liter/mole s, k_4 is 41 liter/mole s; if k_2 is 14×10^3, k_4 will be 20 liter/mole s.)

The termination rate constant of monomer radical with polymer radical in the particle, k_{80}, can be calculated using eqn. (36). From Table IX, C_2 is $1{\cdot}33 \times 10^6$ mole/liter$^{1/3}$. Since D_0 is $1{\cdot}0 \times 10^{-8}$ dm^2/s, k_{80} is $1{\cdot}07 \times 10^9$ liter/mole s. This is comparable to the termination rate constant for vinyl chloride, 2×10^9 liter/mole s.[20] For vinyl acetate, Schulz and Slein found that the termination rate constant ranged from 4 to 8×10^8 liter/mole s.[33] This is in reasonable agreement with the present value.

Estimation of rate constant for induced decomposition of initiator by monomer radical in the aqueous phase, k_7

From eqn. (34), we know that the rate of polymerization is proportional to $(I)^{1/2}(1 + C_3(I)(M)^{1/2})^{1/2}$. Since the experimental data show that polymerization rate is proportional to initiator concentration to the 0·6 power, the additional 0·1 power must come from the term $C_3(I)(M)^{1/2}$. A reasonable approximation is to say that $C_3(I)(M)^{1/2}$ averages 0·2 over the polymerization conditions. Then the square root of $(1 + C_3(I)(M)^{1/2})$ is 1·1 which will contribute approximately 0·1 power on initiator concentration. Since the initiator concentration is changing during the polymerization due to induced decomposition by monomer radical, discussed later,[34] the initial value of initiator cannot be used. The rate of polymerization was taken from the slope of the conversion versus time plot from 30% to 90% conversion. Since the initiator concentration is also a function of conversion,[34] we took the initiator concentration near the mid range of the linear portion of the curve, at 50% conversion, for the calculation. As the polymerization rate is proportional to initiator concentration to $0{\cdot}60 \pm 0{\cdot}03$, $C_3(I)(M)^{1/2}$ may vary from 0·16 to 0·33 due to the errors involved.

$$0{\cdot}16 \leqslant C_3(I)(M)^{1/2} = \frac{0{\cdot}072 k_7 D_0 (M)^{1/2} (I)}{k_4 D_a d^3 (P) P_e (M_0)^2} \leqslant 0{\cdot}33 \qquad (40)$$

As all constants in eqn. (40) except k_7 are known, it can be calculated. Table XI (p. 129) lists the set of constants and parameter values used to calculate k_7. Since $C_3(I)(M)^{1/2}$ can range from 0·16 to 0·33, and (I_0) varies between 1 to 10×10^{-3} M, k_7 is estimated to lie between 2×10^7 and 2×10^8 liter/mole s. The value is quite high, but we have further evidence that justifies it. This will be discussed fully later.[34] Since k_7 is calculated from a minor term, it cannot be estimated accurately here. It is about 6×10^7 liter/mole s within a factor of 3.

Calculation of relative amount of Smith–Ewart termination

While the Smith–Ewart termination was included as a perturbation in our derivation of the rate expression for vinyl acetate emulsion polymerization, we ignored it in the calculations. Here we will verify that this term is much less than 1, based on the rate constants that we obtained. The term is

$$\frac{fk_1(I)\left((1-P_e) + \dfrac{k_7(I)}{2\pi D_a d(P)}\right)^{1/2}}{k_3 P_t P_e (M)(P)}$$

The constants in this term are now all known. P_e and P_t can be calculated, eqns. (15) and (19). We use $k_7 = 2{\cdot}0 \times 10^7$ liter/mole s, the more accurate value determined from induced decomposition of initiator.[34]

At 10% conversion, the highest value for the Smith–Ewart termination term is 3·7% and the lowest is 0·16%. At 90% conversion, the lowest value for Smith–Ewart termination is 0·7%. For most cases, the value ranges from 0·7% to 12%; the highest value is 15% at $(I_0) = 10 \times 10^{-4}$ M and $(P) = 1{\cdot}04 \times 10^{17}$/liter H_2O. But even in this case, the Smith–Ewart termination step is still a minor factor.

Consequences of probability of termination (P_t) being less than 1

The effectiveness of monomer radical-growing radical ($M_p \cdot P \cdot$) termination also influences the kinetics. In final form, $1/P_t$ is written as

$$\frac{1}{P_t} = 1 + \frac{1{\cdot}42 \times 10^{23} D_0 d}{k_{80}(M)^{1/2}} \tag{41}$$

The major change due to the explicit inclusion of P_t is a further dependence on d. This appears in eqn. (33) and Table X as a lower dependence on (P) and a higher dependence on V_{org}/V_{aq} than the official ones of $(P)^{1/6}$ and $(V_{org}/V_{aq})^{0{\cdot}33}$.

There is a further minor effect, which is obvious when theoretical curves are compared, but obscure in the actual data. Since $P_t^{-1/2}$ is proportional to a negative power of (M) ($-0{\cdot}05$ to $-0{\cdot}20$, depending on the value of P_t) this lowers the overall dependence of (33) on (M) from 0·25 to 0·2 or even 0·05 power, and thus changes the curve shape. Under conditions where P_t is small (large particle size due to large V_{org} or small (P)), the curves are linear over a larger range than when $P_t \approx 1$. Thus at large d the integrated equations do not match the data well unless P_t is included.

If one estimates the value of k_{80} necessary for P_t to be $\geqslant 0{\cdot}9$, it would be about 10^{10}. This is obviously very high, even for small radical-small radical termination in a liquid. Thus, P_t must be less than 1.

The next section concerns the effect of temperature, which affects the relationship between D and (M). It should be remembered that lowering P_t changes the polymerization curve shape, lowering its dependence on (M). This helps to produce the effects discussed next.

Effect of temperature on the shape of the polymerization curve

As was mentioned earlier, $D^{1/2}$ is proportional to (M) when the temperature is about 15–30°C above the Tg of the polymer. As can be seen from Figs. 10 and 11, when the temperature is lower, the decrease of $D^{1/2}$ is faster than that of (M). This implies that the order of the dependence of the diffusion coefficient on monomer concentration in monomer–polymer system is more than 2. From eqn. (31) we know the polymerization rate is proportional to $(M)^{5/4}/D^{1/2}(1 + 1{\cdot}42 \times 10^{23} D_0 d/k_{80}(M)^{1/2})^{1/2}$. When $D^{1/2}$ is proportional to (M) to greater than the first power and P_t is low, the polymerization rate can become inversely proportional to monomer concentration to a small power, and the polymerization rate will accelerate as monomer concentration becomes smaller (high conversion). In Friis' work,[1] he used 50°C for most of the emulsion polymerization of vinyl acetate; at low initiator concentrations, acceleration is seen when the conversion is above 60%. As was shown in Fig. 12, D is proportional to $(M)^{2{\cdot}1}$ at 50°C for vinyl acetate in poly(vinyl acetate). Most of our polymerizations are run at 60°C where D is about proportional to $(M)^{2{\cdot}0}$ and there is no acceleration. Figure 17 which is taken

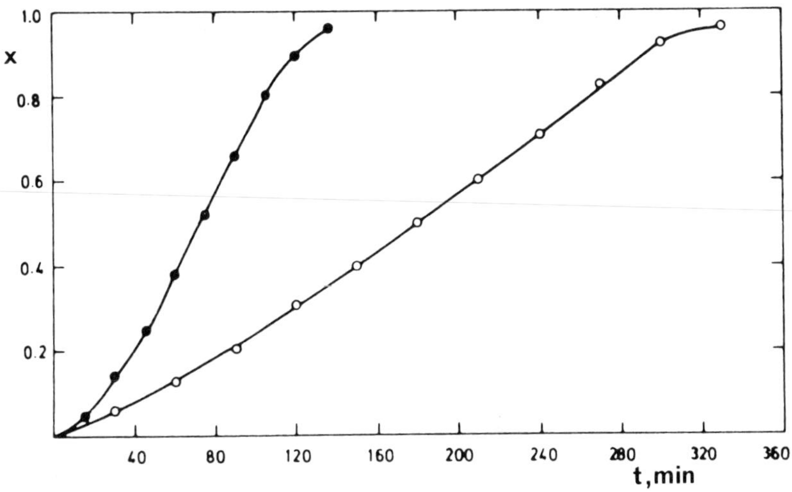

Fig. 17. Conversion versus time plots at two different temperatures. ● 50°C, ○ 60°C. 9·5 g SLS/liter H_2O, 5×10^{-4} moles $K_2S_2O_8$/liter H_2O. (From Friis[1]).

from Friis' thesis[1b] clearly shows that acceleration can occur at 50°C under conditions where it does not at 60°C.

The order of the dependence of D on (M) should be larger at lower temperature. Thus a polymerization was run at 40°C to emphasize the acceleration of the rate. The data are shown in Fig. 18. Acceleration is seen after 50% conversion. It must be emphasized that acceleration is found here under conditions where about 1% of the particles contain radicals. It is not the gel effect. It is primarily because as $D^{1/2}$ decreases faster than (M), the concentration of growing radicals increases faster with conversion than (M) decreases, raising the rate of polymerization.

Comparison with previous models

Our model is different from previous models in that we have explicitly described the emulsion polymerization of vinyl acetate at the molecular level. Each kinetic step is defined exactly and the rate constants are reasonable and can be justified. Chain transfer to the monomer is proposed to be on the vinyl hydrogen of the vinyl acetate, which other theories have not considered. The termination step is the reaction of monomer radical with polymer radical in the particle and its kinetics are also described. The present theory can fit exactly the order of dependence of polymerization rate on the various parameters when the probability of termination is considered explicitly. Other

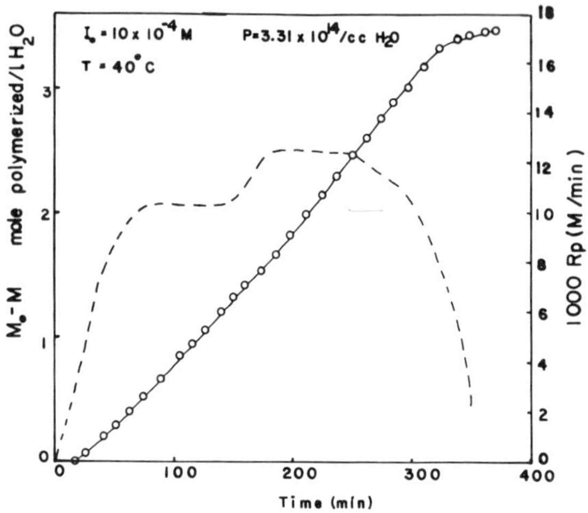

Fig. 18. The polymerization curves (—o—o—o—) and the rate of polymerization (— — —) at 40°C (V_{org}/V_{aq} = 0·33, $\mu = 3 \times 10^{-3}$, soap = Polystep B-19).

theories cannot generate these exactly, though they reproduce the major dependencies. At lower polymerization temperatures, the acceleration of the polymerization rate after 50% conversion can be understood, based on the change of the diffusion coefficient of monomer radical in the particle relative to monomer concentration.

Overview

Our experiments and model have considered most of the possible parameters, i.e. types of soap, concentration of soap, ionic strength, initiator concentration, monomer volume, and particle concentration, and obtained a very good fit between theory and the experimental data. The main features of our theory are: first, chain transfer to monomer generates a stable radical which reinitiates relatively slowly. Monomer radical can then diffuse easily out of particles. These monomer radicals can diffuse through several particles until they either reinitiate or terminate polymerization. Second, the square root of the diffusion coefficient of monomer radical in the particle is approximately proportional to monomer concentration to the first power at 60°C, and to a higher power at lower temperatures. Because of this, the effect of lowering the polymerization temperature is to accelerate the rate at higher conversions. Third, the Smith-Ewart termination term has been shown to be unimportant compared to the major termination step, which is the reaction of monomer radical with polymer radical in a particle. Fourth, the recognition that the probability of termination need not be unity produces exact agreement between theory and experiment over the whole parameter range.

The combination of these features describes in detail vinyl acetate emulsion polymerization. The kinetic steps in our model are the elementary chemical reactions and therefore picture quite clearly the activities of the radical species involved. The steps are either well accepted, or are justified by literature references or internal data. The major controversial step, chain transfer to vinyl, is fully justified later.

SUMMARY AND CONCLUSIONS

All the elementary equations in our proposed model are well defined and are normal reactions of vinyl acetate and initiator. The kinetics are modified because we are operating in a stable two-phase system. Chain transfer to monomer is important and generates monomer radicals which can escape from the particles. Induced decomposition occurs due to the attack on persulfate by monomer radical, which causes an additional 0·1 power dependence of polymerization rate on initiator concentration.

The calculated k_4 value (29 liter/mole s) shows that the monomer radical is quite stable, so it can diffuse easily out of the primary particle. The probability of escape from smaller particles is greater than 80%, Table XI. As chain

TABLE XI
Constants used for calculation of k_7

(I) (mole/liter)[a]	$2{\cdot}2 \times 10^{-4}$
D_0 (dm^2/s)[b]	$1{\cdot}0 \times 10^{-8}$
D_a[35] (dm^2/s)	$2{\cdot}0 \times 10^{-7}$
k_4 (liter/mole s)	29
d (cm)	$1{\cdot}17 \times 10^{-5}$
(P) (/liter)	$3{\cdot}31 \times 10^{17}$
(M) (mole/liter)[c]	5
(M_0) (mole/liter)	7
P_e[d]	0·82

[a] The initiator concentration is taken as its value at 50% conversion. (I_0) was $3{\cdot}0 \times 10^{-4}$ M.
[b] The value is taken from Friis and Nyhagen.[1a]
[c] Monomer concentration at 50% conversion.
[d] Probability of escape at 50% conversion (calculated using eqn. (15)).

transfer on the acetyl group gives a reactive radical, the slow reinitiation implies that a vinyl radical is produced.

The square root of the diffusion constant of monomer radical in the particle at 60°C is suggested to be proportional to the monomer concentration. This gives the result that polymerization rate is approximately constant from 30% to 90% conversion. The major source of monomer dependence is from the partition coefficient. At lower temperatures (40°C and 50°C) the diffusion constant is proportional to monomer concentration to a higher power than at 60°C. This can result in acceleration of the polymerization rate after 50% conversion, even though only a few percent of the particles contain radicals. The model explains this readily. With conversion, radical concentration rises more rapidly than the monomer concentration declines, thus raising the rate of polymerization.

The probability of termination of monomer and growing radicals in the same particle is an important factor in interpreting the experimental results. For large particles the probability of termination is smaller than for the smaller particles. Generally, the dependence of rate on monomer concentration is lowest when the particles are largest, which happens in this theory when the probability of termination is relatively low. Also, the order of dependence of polymerization rate on particle concentration and volume of organic phase is correlated exactly only by including the probability of termination explicitly.

The fact that the polymerization rate is essentially independent of soap concentration and ionic strength implies that escape or resorption of the monomer radical is not affected by the soap concentration and ionic strength in the emulsion. This also implies that no ionic radicals except persulfate are important.

The Smith-Ewart termination term was calculated to be under 12% up to 90% conversion in all cases except one. But at worst it is still a minor term which affects the rate only slightly even above 90% conversion. The general good agreement shows that the neglect is justified.

APPENDIX A

CALCULATION OF DIFFUSION CONSTANT OF A SMALL MOLECULE OUT OF A PARTICLE

The two cases we consider are: (1) when a radical is generated anywhere in the particle, and (2) when a radical enters at the edge of the particle. Referring to Fig. A1, a radical is generated at a distance, y, from the center of a

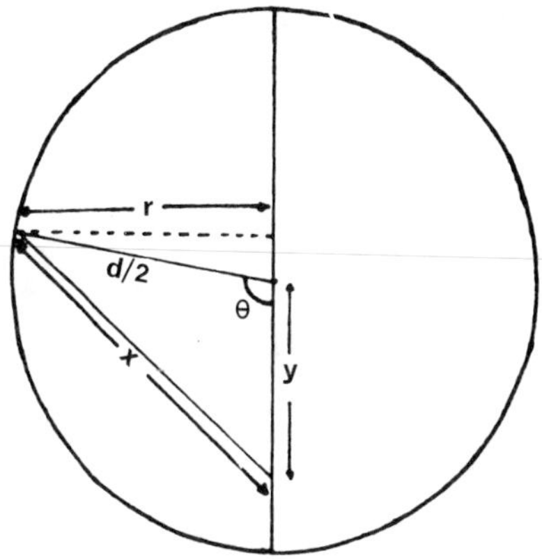

Fig. A1.

particle with radius $d/2$. It travels a distance, x, to escape and it can escape anywhere at the surface of the particle. It can be generated in any volume element within the particle; this is a function of y. The diffusion equation is $x^2 = 2Dt$, where D is the diffusion constant of the radical.

The average time spent by the radical in the particle is then:

$$\bar{t} = \frac{1/(2D) \int_0^{d/2} \int_0^{\pi} x^2 f(\theta) f(y) \, d\theta \, dy}{\int_0^{d/2} \int_0^{\pi} f(\theta) f(y) \, d\theta \, dy} \tag{A1}$$

$$f(\theta) = 2\pi r = 2\pi(d/2)\sin\theta \tag{A2}$$

$$f(y) = 4\pi y^2 \tag{A3}$$

$$x^2 = (d/2)^2 + y^2 - 2(yd/2)\cos\theta \tag{A4}$$

Substituting (A2), (A3) and (A4) into (A1) and integrating we find:

$$\bar{t} = d^2/5D \tag{A5}$$

For special case 1, where the radical has just entered the particle, y is fixed at $d/2$. In such a situation, $\bar{t} = d^2/4D$.

The difference between the two cases is small and will be insignificant for the kinetic analysis. Since we postulate that the radical diffuses through many particles we will consider that for all cases $\bar{t} = d^2/4D$. The rate constant for diffusion out of a particle is $1/\bar{t}$ and is then:

$$1/\bar{t} = 4D/d^2$$

APPENDIX B

INTEGRATION OF EQN. (34)

$$Rp = C_2(1 + C_2 d/(M)^{1/2})^{1/2} (M)^{1/4} (V_{org}/V_{aq})^{1/3} (P)^{1/6} (I)^{1/2} (1 + C_3(I)(M)^{1/2})^{1/2} \tag{34}$$

Polymerization in the presence of separate monomer phase

The polymerization can be divided into two sections as discussed in the text. In the first section, between 0 and 30% conversion when a separate monomer phase is present, (M), (P) and C_1 are constant during polymerization. Initiator concentration is changing due to the induced decomposition of initiator by monomer radical in aqueous phase.[20] For simplicity, an average (I) is taken for the integration. $C_2 d$ and C_3 also change as the polymerization proceeds from 0 to 30% conversion. Since they are minor terms, we make the approximation that they are constant during the polymerization and take their average value. Now the only variable is (V_{org}/V_{aq}) as the particles grow. Equation (34) can be written as

$$\frac{-\mathrm{d}(M')}{\mathrm{d}t} = k_1' y \tag{B1}$$

where $k_1' = C_1(1 + C_2 d/(M)^{1/2})^{1/2} (M)^{1/4} (P)^{1/6} (I)^{1/2} (1 + C_3(I)(M)^{1/2})^{1/2}$ and

$$y = \left(\frac{V_{org}/V_{aq}}{V_{orgf}/V_{aq}}\right)^{1/3} = (V_{org}/V_{orgf})^{1/3}$$

V_{orgf} is the final volume of organic phase. Thus

$$y^3 = \frac{V_{org}}{V_{orgf}} = \frac{(M_0') - (M')}{b(M_0')} \tag{B2}$$

where (M_0') is the initial monomer plus polymer concentration per liter H_2O in the emulsion, (M') is the monomer concentration during polymerization, b

is the volume fraction of polymer in the particles in equilibrium with a separate monomer phase (usually $b = 0{\cdot}30$). From eqn. (B2), eqn. (B3) is obtained

$$\frac{-\mathrm{d}(M')}{\mathrm{d}t} = (M_0')\, b\, \frac{\mathrm{d}y^3}{\mathrm{d}t} = 3(M_0')\, by^2\, \frac{\mathrm{d}y}{\mathrm{d}t} \tag{B3}$$

Substituting (B3) into (B1), (B4) is obtained.

$$k_1' y = 3(M_0')\, by^2\, \frac{\mathrm{d}y}{\mathrm{d}t} \tag{B4}$$

Integration of (B4) from $y = y_\mathrm{i}$ to y gives

$$y^2 - y_\mathrm{i}^2 = \frac{2k_1' t}{3(M_0')\, b} \tag{B5}$$

where $y_\mathrm{i} = (V_{\mathrm{org_i}}/V_{\mathrm{org_f}})^{1/3}$, where $V_{\mathrm{org_i}}$ is the volume of swollen seeds and $V_{\mathrm{org_f}}$ the volume of final organic phase. The final equation after integration and replacing y by a function of $[M']$ is

$$\left(\frac{(M_0') - (M')}{b(M_0')}\right)^{2/3} - \left(\frac{V_{\mathrm{org_i}}}{V_{\mathrm{org_f}}}\right)^{2/3} = \frac{2C_1(1 + C_2 d/(M_0)^{1/2})^{1/2}(M_0)^{1/4}(P)^{1/6}(I)^{1/2}(1 + C_3(I)(M_0)^{1/2})^{1/2} t}{3(M_0')\, b} \tag{B6}$$

Polymerization after separate monomer phase is consumed (about 30% conversion)

During the second part of polymerization, the particle volume remains about constant, the particle concentration (P), and C_1 are constants. $C_2 d$ is constant after the separate monomer phase disappears. Initiator concentration, (I), is changing slightly and the average initiator concentration from 30 to 90% conversion is taken during integration. $C_3(M)^{1/2}$ is approximately independent of monomer concentration since the probability of escape (term included in C_3) also changes as a function of monomer concentration. Now the monomer concentration in the organic phase, $[M]$, is decreasing.

Thus, (34) can be written as

$$Rp = \frac{-\mathrm{d}(M)}{\mathrm{d}t} = k_2'(M)^{1/4}(1 + C_2 d/(M)^{1/2})^{1/2} \tag{B7}$$

where $k_2' = C_1(V_{org}/V_{aq})^{1/3}(P)^{1/6}(I)^{1/2}(1 + C_3(I)(M)^{1/2})^{1/2}$. (B7) can be integrated after substituting (B8)

$$(M') = \frac{V_{orgf}}{V_{aq}} (M) \tag{B8}$$

into (B7) and dividing both sides by (M_0'), (B9).

$$\frac{-\mathrm{d}\left(\frac{(M')}{(M_0')}\right)}{\mathrm{d}t} = \frac{k_2'}{(V_{org}/V_{aq})^{1/4}} \frac{(M')^{1/4}}{(M_0')} \left(1 + \frac{\frac{C_2 d}{(M_0')^{1/2}}\left(\frac{V_{org}}{V_{aq}}\right)^{1/2}}{((M')/(M_0'))^{1/2}}\right)^{1/2} \tag{B9}$$

Let

$$x^2 = (M')/(M_0') \tag{B10}$$

Then

$$\mathrm{d}((M')/(M_0')) = 2x\,\mathrm{d}x \tag{B11}$$

Substituting (B11) and (B10) into (B9), and integrating, (B12) is obtained.

$$\left[2x\left(x + \frac{C_2 d\left(\frac{V_{org}}{V_{aq}}\right)^{1/2}}{(M_0')^{1/2}}\right)^{1/2} - \frac{4}{3}\left(x + \frac{C_2 d\left(\frac{V_{org}}{V_{aq}}\right)^{1/2}}{(M_0')^{1/2}}\right)^{3/2}\right]_{x_i}^{x} = -\frac{k_2'(t - t_i)}{2(V_{org}/V_{aq})^{1/4}\cdot (M_0')^{3/4}} \tag{B12}$$

Rearranging (B12) produces (B13)

$$\left(x_i + \frac{C_2 d\left(\frac{V_{org}}{V_{aq}}\right)^{1/2}}{(M_0')^{1/2}}\right)^{1/2}\left(\frac{x_i}{3} - \frac{2}{3}\frac{C_2 d\left(\frac{V_{org}}{V_{aq}}\right)^{1/2}}{(M_0')^{1/2}}\right) - \left(x + \frac{C_2 d\left(\frac{V_{org}}{V_{aq}}\right)^{1/2}}{(M_0')^{1/2}}\right)^{1/2}\left(\frac{x}{3} - \frac{2}{3}\frac{C_2 d\left(\frac{V_{org}}{V_{aq}}\right)^{1/2}}{(M_0')^{1/2}}\right) = \frac{k_2'(t - t_i)}{4(V_{org}/V_{aq})^{1/4}\cdot (M_0')^{3/4}} \tag{B13}$$

Usually, the separate monomer phase disappears at 30% conversion, thus $x_i = ((M')/(M_0'))^{1/2} = (0{\cdot}7)^{1/2} = 0{\cdot}84$. Substituting for x_i into (B13), and replacing k_2' the final equation is (B14).

$$\left(0{\cdot}84 + \frac{C_2 d\left(\dfrac{V_{\text{org}}}{V_{\text{aq}}}\right)^{1/2}}{(M_0')^{1/2}}\right)^{1/2}\left(0{\cdot}28 - \frac{2}{3}\,\frac{C_2 d\left(\dfrac{V_{\text{org}}}{V_{\text{aq}}}\right)^{1/2}}{(M_0')^{1/2}}\right) - \left[\left(\frac{M'}{M_0'}\right)^{1/2} + \frac{\left(\dfrac{V_{\text{org}}}{V_{\text{aq}}}\right)^{1/2}}{(M_0')^{1/2}}\right]\left[\frac{1}{3}\left(\frac{(M')}{(M_0')}\right)^{1/2} - \frac{2}{3}\,\frac{\left(\dfrac{V_{\text{org}}}{V_{\text{aq}}}\right)^{1/2}}{(M_0')^{1/2}}\right]$$

$$= \frac{C_1(V_{\text{org}}/V_{\text{aq}})^{1/12}(P)^{1/6}(I)^{1/2}(1 + C_3(I)(M)^{1/2})^{1/2}(t - t_i)}{4(M_0')^{3/4}} \qquad \text{(B14)}$$

ACKNOWLEDGEMENTS

The authors thank the Polymer Section of the Materials Science Division of National Science Foundation for partial support of this work by Grant No. DMR 76-04562. The authors also want to thank Air Products Corporation for partial support of M. Nomura.

REFERENCES

1. (a) N. Friis and L. J. Nyhagen, *J. Appl. Polym. Sci.*, **17**, 2311 (1973).
   (b) N. Friis, Thesis, Technical University, Copenhagen, Denmark, 1973.
2. (a) M. Nomura, M. Harada, K. Nakagawara, W. Eguchi and S. Nagata, *J. Chem. Eng. Jpn.*, **4**, 160 (1971).
   (b) M. Nomura, M. Harada, W. Eguchi and S. Nagata, *ACS Symposium Series*, No. 24, 1976, p. 102.
3. R. Patsiga, Thesis, New York State University, College of Forestry, Syracuse, New York, 1962.
4. M. H. Litt, R. Patsiga and. V. T. Stannett, *J. Polym. Sci., A1*, **8**, 3607 (1970).
5. D. Gershberg, Paper Presented at Joint Meeting of AIChE and IChE (England), London, June 14, 1965.
6. J. W. Breitenbach, H. Edelhauser and R. Hochrainer, *Monatshefre fur Chemie*, **99**, 625 (1968).
7. A. S. Dunn and P. A. Taylor, *Makromol. Chem.*, **83**, 207 (1965).
8. A. S. Dunn and L. C.-H. Chong, *Br. Polym. J.*, **2**, 49 (1970).
9. W. V. Smith and R. W. Ewart, *J. Chem. Phys.*, **16**, 592 (1948).

10. V. T. Stannett, A. Klein and M. H. Litt, *Br. Polym. J.*, **7**, 139 (1975).
11. P. Harriot, *J. Polym. Sci., A-1,* **9**, 1153 (1971).
12. D. M. French, *J. Polym. Sci.,* **32**, 395 (1958).
13. M. Lazar, J. Paulinec and Z. Manasek, *Collection Czech. Chem. Commun.,* **26**, 1380 (1961).
14. S. Imoto, J. Ukida and T. Kominami, *Kobunshi. Kagaku,* **14**, 101 (1957).
15. O. L. Wheeler, E. Lavin and R. N. Crozier, *J. Polym. Sci.,* **9**, 157 (1952).
16. R. O. Howard, Ph.D. Thesis, Massachusetts Institute of Technology, Cambridge, Mass., 1952.
17. P. K. Isaacs and H. A. Edelhauser, *J. Appl. Polym. Sci.,* **10**, 171 (1966).
18. K. H. S. Chang, M. H. Litt and A. M. Jamieson, The measurement of latex particle size by quasielastic laser light scattering, *J. Polym. Sci., A-2,* submitted for publication.
19. M. L. Corrin, *J. Polym. Sci.,* **2**, 257 (1947).
20. G. M. Burnett and W. W. Wright, *Proc. Roy. Soc. (London),* **A211**, 41 (1954).
21. M. H. Litt and K. H. S. Chang, this volume, Chapt. 8.
22. M. Ryska, M. Kolinsky and D. Lim, *J. Polym. Sci., C,* **33**, 357 (1971).
23. J. W. Breitenbach and A. Schindler, *Monatsh.,* **86**, 437 (1955).
24. J. W. Breitenbach, O. F. Olaj, H. Reif and A. Schindler, *Makromol. Chem.,* **122**, 51 (1969).
25. K. Higashiura and M. Oiwa, *J. Polym. Sci., A-1,* **6**, 1857 (1968).
26. K. Goffloo and R. Kosfeld, *Makromol. Chem.,* **37**, 105 (1974).
27. R. Kosfeld and K. Goffloo, *Kolloid Z.U. Polymere,* **247**, 801 (1971).
28. H. Fujita, *Adv. Polym. Sci.,* **3**, 1 (1961).
29. D. H. Napper and A. G. Parts, *J. Polym. Sci.,* **61**, 113 (1962).
30. E. Vanzo, R. H. Marchessault and V. T. Stannett, *J. Colloid Sci.,* **20**, 62 (1965).
31. I. M. Kolthoff and J. K. Miller, *J. Amer. Chem. Soc.,* **73**, 3055 (1951).
32. J. Brandrup and E. H. Immergut (eds.), *Polymer Handbook,* Interscience, New York, 1975, (a) p. II-50, (b) p. II-62.
33. G. V. Schulz and D. J. Slein, *Makromol. Chem.,* **52**, 1 (1962).
34. M. H. Litt and K. H. S. Chang, this volume, Chapt. 7.
35. C. R. Wilke and P. Chang, *AIChE Journal,* **1**, 264 (1955).

7

The Reinvestigation of Vinyl Acetate Emulsion Polymerization (II) – The Induced Decomposition of Initiator

M. H. LITT and K. H. S. CHANG

Department of Macromolecular Science, Case Western Reserve University, Ohio, USA

SUMMARY

In Chapter 6, a model was presented and the kinetic equations developed for vinyl acetate emulsion polymerization. The polymerization rate was found to be dependent on initiator concentration to the 0·6 power. In the model, 0·1 power of the dependence was due to the induced decomposition of the initiator by monomer radical. In this work, the initiation concentration as a function of time was followed as a function of various parameters.

The data show that the change of the natural logarithm of the initiator concentration with time is proportional to the initiator concentration to the 0·6 power, number of particles to the 0·11 power and volume of organic phase per unit volume of aqueous phase to the −0·64 power. All these dependencies found are in very good agreement with those predicted from the theory. Numerical integrations follow the decomposition exactly.

The rate constant for induced decomposition was calculated by three independent methods (four different values). The values obtained from these four calculations show that this constant is in the range of $10^7 \sim 10^8$ *liter/mole s. The best value is* $2{\cdot}0 \times 10^{-7}$ *liter/mole s. This implies that the reaction of persulfate ion with monomer radical is diffusion controlled and probably occurs by an electron transfer mechanism.*

INTRODUCTION

In aqueous solution, potassium persulfate is believed to decompose by the following steps:

(a) $S_2O_8^{-2} \rightarrow 2SO_4^{-}\cdot$

(b) $2SO_4^{-}\cdot + 2H_2O \rightarrow 2HSO_4^{-} + 2HO\cdot$

(c) $2HO\cdot \rightarrow H_2O + \frac{1}{2}O_2$

Step (a) is a symmetrical scission of the O–O bond to form two sulfate ion radicals.[1] In an emulsion polymerization, these radicals can easily enter micelles and particles or add to monomer dissolved in the aqueous phase. Smith[2] and Smith and Campbell[3] found that persulfate initiates polymerization of styrene by the addition of $SO_4^{-}\cdot$ to the styrene in the micelles in emulsion polymerization. Kolthoff and Miller[1] found that the first order rate constant of decomposition for potassium persulfate, is $3{\cdot}0 \times 10^{-4}\,\text{min}^{-1}$ at 60°C in a neutral aqueous medium. Decomposition is more rapid at low pH and is believed to be the result of induced decomposition by hydrogen ion. Also, decomposition is accelerated by many water soluble or partially soluble organic compounds such as methanol and ethanol,[4] which cause a large increase in rate. Bartlett and Nozaki[5] found that ethyl acetate can also increase the rate of decomposition.

In Patsiga's thesis,[6] the decomposition of potassium persulfate was followed in an aqueous 0·67% sodium lauryl sulfate solution of 60°C. He found that the first order rate constant was $5{\cdot}07 \times 10^{-4}\,\text{min}^{-1}$ which is slightly higher than the value obtained by Kolthoff and Miller. This is reasonable since in Patsiga's work soap was present in the solution while Kolthoff and Miller were looking at the neutral aqueous medium.

Patsiga did not measure the decomposition of persulfate as polymerization proceeded, but analyzed the persulfate remaining in the latexes at the end of each run. He found that the decomposition was ten times more rapid in the presence of vinyl acetate than with soap alone or in styrene emulsion polymerization. The values given in Patsiga's thesis for the total fraction of initiator decomposed vary considerably, from 10% to 45%. He felt that this was partly due to the difficulty in extracting all persulfate from the coagulated latex.

We thought that a detailed analysis of persulfate decomposition as a function of conversion and the standard variables, such as organic volume, initiator concentration and particle concentration, would shed further light on the kinetics of vinyl acetate emulsion polymerization as well as the mechanism of induced decomposition. We report on this work below.

EXPERIMENTAL

Materials

Initiator

The initiator, potassium persulfate, was Fisher certified reagent grade and was used directly from the bottle. Potassium sulfate which was added to the polymerization system to maintain a constant ionic strength of 3×10^{-3} was MC/B reagent grade and was used directly.

Monomer

Vinyl acetate was obtained from Eastman Kodak Company and purified by distillation through a Perkin Elmer Spinning Band Column with 45 theoretical plates. The first and the last 25% of the distillate were discarded for all the distillations. The middle cut was then stored at 5°C.

Emulsifier

A purified grade of sodium lauryl sulfate (Stepanol WA-100) was obtained from Stepan Chemical Company and was used directly. A purified grade of sodium lauryl ether sulfate (Polystep B-19, 30 ethylene oxide units), 20 wt% in aqueous solution was also obtained from Stepan Chemical Company and was used directly from the bottle.

Nitrogen

The nitrogen was obtained from Matheson Gas Products (research purity, purity $>$ 99·9995%, $O_2 < 1$ ppm).

Preparation of seed latex

The procedure and the recipes for the preparation of the seed latex were given in Chapter 6. The seed latex used had a solids content of 17% and its particle number average diameter was determined as 0·045 μm by quasi-elastic laser light scattering.[7]

Apparatus and polymerization procedure

The design of the experimental apparatus is the same as was shown in Fig. 1 of Chapter 6. The polymerization procedure is the same as that for the preparation of seed latex (Chapter 6) except that samples were withdrawn during polymerization and the final latex had 25% polymer. Once the polymerization started, 15 cc samples were taken every five to ten minutes. About 5 cc of each sample was used to determine the monomer conversion gravimetrically. Weighed samples of the latexes were dried in weighing bottles at 80°C until they reached constant weight. This generally took less than 48 hours. The conversion was calculated, subtracting the weight of soap in the latex sample. The remainder of each sample was used to measure the residual initiator concentration.

Experiments were run varying the following parameters: volume of organic phase per unit volume of aqueous phase, V_{org}/V_{aq}, particle concentration,

TABLE I
Experimental conditions for the study of persulfate decomposition

Run No.	V_{org}/V_{aq}	$(P) \times 10^{-14}$ (cc^{-1})	$(I_0) \times 10^4$ (mole/liter)	*Emulsifier*
114	0·33	7·27	5	SLS[a]
114A	0·33	7·27	5	SLES[b]
115	0·33	7·27	10	SLS
116	0·20	7·27	5	SLS
117	0·46	7·27	5	SLS
118	0·20	3·63	5	SLS
119	0·20	2·42	5	SLS
120	0·33	7·27	2	SLS
121	0·20	1·45	5	SLS
122	0·33	7·27	3	SLS

[a] Sodium lauryl sulfate (Stepanol WA-100).
[b] Sodium lauryl ether sulfate (Polystep B-19).

(P), and initiator concentration, (I). The exact values for the variables are given in Table I.

Determination of persulfate concentration

Generally ten latex samples were withdrawn per run. A typical analytical procedure is given below. It was modified from that used by Patsiga[6] to circumvent the difficulties he encountered. The sample was cooled in ice water and 10 cc was centrifuged using a duPont Sorvall RC5 centrifuge at 16 000 rpm (equivalent to 30 877 gravities) at 4°C for one hour. Two separate phases were obtained; the upper phase was a clear aqueous phase while the lower phase was compacted polymer. To 1 cc of the aqueous phase was added 0·5 cc of a solution 0·01 molar in ferrous ammonium sulfate and 1·5 molar in potassium thiocyanate. This was then diluted to exactly 10 cc with 0·01 normal nitric acid. A blank solution was prepared by adding 0·5 cc of a 1·5 molar solution of potassium thiocyanate to 1 cc of the aqueous phase. This was then diluted with 0·01 normal nitric acid to 10 ml. The optical absorption at 470 nm was then measured using a Beckman Acta MII Spectrophotometer. The analysis is based on the color produced when persulfate oxidizes ferrous ions to ferric in the presence of thiocyanate. The reactions are illustrated below:

$$S_2O_8^{-2} + 2Fe^{+2} \rightarrow 2SO_4^{-2} + 2Fe^{+3}$$

$$Fe^{+3} + 6SCN^- \leftrightarrows Fe(SCN)_6^{-3}$$

The ferric thiocyanate complex is deep red in solution and shows a maximum absorption at 470 nm.

For calibration, samples of known persulfate concentration were analyzed and the resultant curve plotted. The $Fe(SCN)_6^{-3}$ absorption followed Beer's law with a molar extinction coefficient of $0{\cdot}66 \times 10^4$. From the optical density of the $Fe(SCN)_6^{-3}$ complex and the molar extinction coefficient, we can calculate the concentration of complex. The persulfate concentration is then known since its concentration is one half that of the complex.

THEORY

The rate of change of initiator concentration as a function of time

As presented in Chapter 6, the growing radical in a particle can chain transfer to monomer, generating a monomer radical. This monomer radical reinitiates relatively slowly so it has time to diffuse out of the particle into the aqueous phase. The aqueous monomer radical can be swept up by another particle, or attack persulfate to generate sulfate and an ion radical, $SO_4^{-}\cdot$. (All symbols are defined in the Nomenclature to Chapter 6.) We will concentrate on the induced decomposition step in this paper.

Based on the proposed kinetic scheme (shown in Chapter 6, page 110) the expression for the change of the initiator concentration with time is derived in Appendix A. The equation for the change of the natural logarithm of initiator concentration as a function of time, Appendix A (A11), is reproduced here, eqn. (1)

$$\frac{-\mathrm{d}\ln(I)}{\mathrm{d}t} = k_1 + 2{\cdot}75 \times 10^{-9} k_7 (fk_1k_3/k_4)^{1/2} (D_0^{1/2}/(M_0)) \left(\frac{1}{D_a}\right) (M)^{3/4} \times \left(1 + \frac{1{\cdot}422 \times 10^{23} D_0 d}{k_{80}(M)^{1/2}}\right)^{1/2} (P)^{1/6} (V_{org}/V_{aq})^{-2/3} (I)^{0{\cdot}6} \qquad (1)$$

Initiator concentration as a function of time

Referring to eqn. (1), k_1 is the first order rate constant for persulfate decomposition. The group constant, $2{\cdot}75 \times 10^{-9} k_7(fk_1k_3/k_4)^{1/2}$ is a combination of rate constants which arises from the induced decomposition of persulfate by monomer radical. (For convenience it will be written as K_2.) Integration of (1) must be separated into two parts because the variables change depending on whether the reaction is below or above the conversion at which the separate monomer phase disappears. The particle concentration, (P), is constant throughout the polymerization since we used a seeded polymerization. But the particle size does change. Below 30% conversion (including the initial seed), the particles grow since there is still a separate monomer phase from which monomer diffuses into the particle and polymerizes there. Only the total particle volume, V_{org}, is considered active. The separate monomer phase is kinetically inert. Thus, the ratio of volume of organic phase to the volume of aqueous phase, (V_{org}/V_{aq}), increases as the polymerization proceeds. V_{org} reaches its final volume, V_{orgf}, when the monomer phase disappears at about 30% conversion. The monomer concentration in the particles, (M), is constant at 7 M below 30% conversion since the separate monomer phase serves as a reservoir. The probability of termination is changing also, but we take the average value between 0 and 30% conversion for this stage of integration of eqn. (1). Above 30% conversion, the monomer concentration, (M), decreases. The probability of termination also changes; we take the average value between 30 to 90% conversion for the integration of eqn. (1) in the second stage.

The expressions for $\mathrm{d}\ln(I)/\mathrm{d}t$ are derived in Appendix B. Below 30% conversion, the expression for change of initiator concentration as a function of time, Appendix B, eqn. (B11), is given as eqn. (2)

$$\frac{-\mathrm{d}\ln(I)}{\mathrm{d}t} = k_1 + \left(\frac{2{\cdot}75\times10^{-9}k_7(fk_1k_3/k_4)^{1/2}(D_0^{1/2}/(M_0))\left(\frac{1}{D_a}\right)((M_0)^{3/4})}{(V_{org_0}/V_{orgf})^{2/3}\ ((M_0')/2{\cdot}22K_3)+t}\right)$$
$$\left(\frac{(1+1{\cdot}422\times10^{23}D_0d/k_{80}(M_0)^{1/2})^{1/2}((P)^{1/6}(M_0')/2{\cdot}22K_3)(I)^{0{\cdot}6}}{(V_{orgf}/V_{aq})^{2/3}}\right) \quad (2)$$

After 30% conversion, the expression for change of initiator concentration as a function of time, Appendix B, eqn. (B15), is given below, eqn. (3). Monomer concentration has been replaced by a function of time.

$$\frac{-\mathrm{d}\ln(I)}{\mathrm{d}t} = k_1 + 2{\cdot}75\times10^{-9}k_7(fk_1k_3/k_4)^{1/2}\ (D_0^{1/2}/(M_0)^{1/4})\left(\frac{1}{D_a}\right)$$
$$\left(1+\frac{1{\cdot}422\times10^{23}D_0d}{k_{80}(M)^{1/2}}\right)^{1/2}(P)^{1/6}\,(V_{org}/V_{aq})^{-2/3}\,(I)^{0{\cdot}6}$$
$$-8{\cdot}48\times10^{-17}fk_1k_2k_7\left(\frac{1}{D_a}\right)\left(1+\frac{1{\cdot}422\times10^{23}D_0d}{k_{80}(M)^{1/2}}\right)^{1/2}$$
$$(V_{org}/V_{aq})^{-1/3}\,(P)^{1/3}\,(I)^{1{\cdot}2}\,(t-t_0) \quad (3)$$

To determine theoretically the initiator concentration as a function of time, we start from the value for initiator concentration at 30% conversion and numerically integrate backward to zero time using eqn. (2), adjusting the group constant $K_2 = 2{\cdot}75 \times 10^{-9} k_7 (f k_1 k_3/k_4)^{1/2}$ until the best fit with data is obtained. Again, using the initiator concentration at 30% conversion, eqn. (3) is numerically integrated forward to the end of polymerization, adjusting K_2 and a second group constant, $8{\cdot}48 \times 10^{-17} f k_1 k_2 k_7$, until the best fit with data is obtained.

RESULTS

As an example of the way in which initiator decomposes as a function of conversion, data for run 117 are shown in Fig. 1. Table II lists the constants that were used to integrate the initiator concentration as a function of time for the different polymerization conditions. As the unimolecular decomposition constant is small compared to the vinyl acetate induced decomposition, the literature value was used.[1]

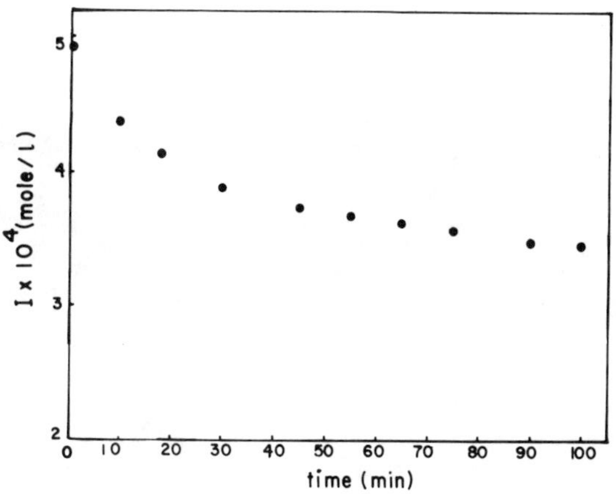

Fig. 1. Initiator concentrations as a function of time for run 117.

As can be seen, the group constant, K_2, remains constant within experimental error for all runs. The average value for K_2 used to integrate below 30% conversion for different runs is $4{\cdot}30 \pm 0{\cdot}08 \times 10^{-5}$ liter/mole $s^{3/2}$; above 30% conversion it is $4{\cdot}22 \pm 0{\cdot}04 \times 10^{-5}$ liter/mole $s^{3/2}$. The average value for all the K_2s is then $4{\cdot}26 \pm 0{\cdot}06 \times 10^{-5}$ liter/mole $s^{3/2}$. The other group constant, $f k_1 k_2 k_7$, averages $1{\cdot}55 \pm 0{\cdot}09 \times 10^{-6}$ (liter/mole)^2/s^3.

TABLE II
The constants used to integrate the initiator concentration as a function of time for different polymerization conditions

	Run No.									
	114	114A	115	116	117	118	119	120	121	122
$k_1 \times 10^4$ (s^{-1})[a]	8·3	8·3	8·3	8·3	8·3	8·3	8·3	8·3	8·3	8·3
Below 30% conversion										
b[b]	7·23	7·34	5·22	7·96	7·00	7·30	7·25	12·15	7·27	9·35
c[c]	8·04	7·86	5·61	8·34	7·67	5·57	4·47	13·05	3·37	10·40
$K_2 \times 10^5$[d]	4·43	4·25	4·35	4·46	4·35	4·24	4·23	4·20	4·29	4·16
Above 30% conversion										
d[e]	0·202	0·207	0·205	0·290	0·165	0·255	0·242	0·205	0·234	0·203
$10^2 e$[f]	5·76	6·00	5·91	6·43	5·36	5·08	4·44	5·99	3·73	5·74
t_i[g]	30	31	23	30	30	26	27	32	34	34
t_0[h]	9	10	7·5	13	7	9	11	4	10	8
$K_2 \times 10^5$	4·20	4·30	4·26	4·12	4·20	4·22	4·19	4·26	4·27	4·22
$10^{-6} f k_1 k_2 k_7$[i]	1·59	1·66	1·64	1·53	1·59	1·47	1·40	1·66	1·33	1·59

[a] The literature value[1] for first order rate constant, k_1, was used here: $k_1 = 8{\cdot}3 \times 10^{-6}\,\mathrm{s}^{-1}$.

[b] $b = 2{\cdot}75 \times 10^{-9} k_7 (fk_1 k_3/k_4)^{1/2} (D_0^{1/2}/(M_0))(1/D_a)(M)^{3/4}/(V_{\mathrm{orgf}}/V_{\mathrm{aq}})^{2/3}$

$$\times \left(1 + \frac{1{\cdot}422 \times 10^{23} D_0 d}{k_{80}(M)^{1/2}}\right)^{1/2} (P)^{1/6} \left(\frac{(M_0')}{2{\cdot}22\, k_3}\right).$$ Unit: (liter/mole)$^{0{\cdot}5}$

[c] $c = (V_{\mathrm{org}_0}/V_{\mathrm{orgf}})^{2/3}((M_0')/2{\cdot}22\, K_3)$. Unit: min. K_3 is the rate of polymerization of each run at 30% conversion.

[d] $K_2 = 2{\cdot}75 \times 10^{-9} k_7 (fk_1 k_3/k_4)^{1/2}$.

[e] $d = 2{\cdot}75 \times 10^{-9} k_7 (fk_1 k_3/k_4)^{1/2} (D_0^{1/2}/(M_0)^{1/4})(1/D_a)$

$$\times \left(1 + \frac{1{\cdot}422 \times 10^{23} D_0 d}{k_{80}(M)^{1/2}}\right)^{1/2} (P)^{1/6} (V_{\mathrm{org}}/V_{\mathrm{aq}})^{-2/3}.$$ Unit: (liter/mole)$^{0{\cdot}5}$/min.

[f] $$e = 8{\cdot}48 \times 10^{-17} fk_1 k_2 k_7 \left(\frac{1}{D_a}\right) \left(1 + \frac{1{\cdot}422 \times 10^{23} D_0 d}{k_{80}(M)^{1/2}}\right) (V_{\mathrm{org}}/V_{\mathrm{aq}})^{-1/3} (P)^{1/3}.$$ Unit: (liter/mole)/min^2.

[g] t_i is the time at 30% conversion.

[h] t_0 is the time at which the plot of (conversion)$^{3/4}$ versus time intercepts the time axis.

[i] $fk_1 k_2 k_7$ is the units of (liter/mole)^2/s^3.

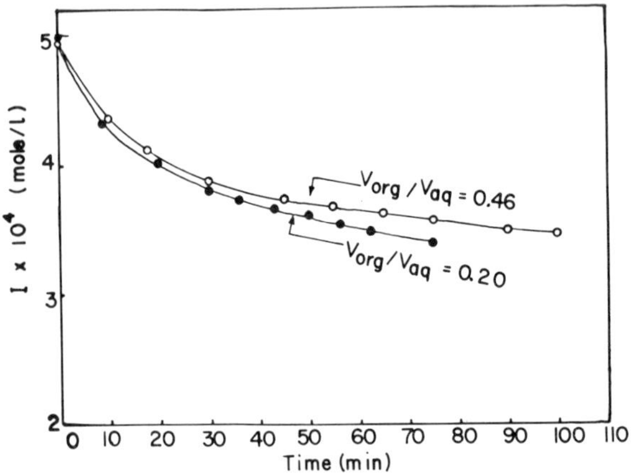

Fig. 2. The integrated curve fitting of initiator concentration as a function of time for runs 116 and 117. ○, ●, Experimental points; ——, theoretical curve.

To show the effect of variation in organic volume, the integrated curves for runs 116 and 117, $(V_{org}/V_{aq}) = 0{\cdot}20$ and $0{\cdot}46$ respectively, are compared with the experimental data in Fig. 2. As expected, as the volume of organic phase decreases, the rate of induced decomposition increases.

DISCUSSION

Comparison of the experimental results with theory

Equation (1) is the expression for the change of the natural logarithm of initiator concentration as a function of time. In order to find the experimental dependence of the rate of induced decomposition of initiator on the various parameters, the values of term d of Table II were plotted against (P) and (V_{org}/V_{aq}) on log–log plots. We know from our curve fitting that the best value for the group constant K_2 is $4{\cdot}26 \times 10^{-5}$ liter/mole $s^{3/2}$. By plugging K_2 and other known values into eqn. (1) the dependence of $(-\mathrm{d}\ln(I)/\mathrm{d}t)$ on (V_{org}/V_{aq}) and (P) can be calculated directly. Table III summarizes the comparison of the theoretical and experimental order of dependence of rate of induced decomposition on the various parameters. As can be seen, the theoretical predictions and experimental results agree quite well.

The rate constant for induced decomposition of persulfate

It was shown in Chapter 6 that the rate constant for induced decomposition of potassium persulfate by the monomer radical in the aqueous phase was in

TABLE III
A comparison of the theoretical and experimental order of dependence of rate of induced decomposition on various parameters

Parameter	*Observed order of dependence*	*Theoretical order of dependence*
(I)	$0{\cdot}60^{a}$	0·60
(P)	0·11 ± 0·01	0·104 ± 0·001
V_{org}/V_{aq}	−0·64 ± 0·01	−0·608 ± 0·005

[a] Assumed value.

the range of $2 \sim 12 \times 10^7$ liter/mole s based on the order of dependence of the polymerization rate on initiator concentration. The value for this rate constant is quite high. Here we give three other independent calculations of this rate constant to see whether this high value is reasonable.

Calculation of induced decomposition rate constant, k_7, from K_2

We have shown in Table III that the group constant K_2 is constant within experimental error for all runs.

$$K_2 = 2{\cdot}75 \times 10^{-9} k_7 (fk_1 k_3/k_4)^{1/2} \tag{4}$$

K_2 ranges from $4{\cdot}12$ to $4{\cdot}46 \times 10^{-5}$ liter/mole $s^{3/2}$. Every term in eqn. (4) is known except k_7 which can be calculated. The unimolecular persulfate decomposition, k_1, is known to be $8{\cdot}3 \times 10^{-6}\ s^{-1}$.[1] The chain transfer constant of vinyl acetate radical to monomer, k_3, is 2·0 liter/mole s.[8b] The reinitiation rate constant by chain transferred monomer radical adding to monomer, k_4, was found to be 29 liter/mole s in Chapter 6. k_7 is calculated to be $2{\cdot}1 \pm 0{\cdot}05 \times 10^7$ liter/mole s if $f = 1{\cdot}0$.

Calculation of k_7 from group constant $fk_1k_2k_7$

From Table III we see that this group constant is relatively constant. $fk_1k_2k_7$ ranges from $1{\cdot}33$ to $1{\cdot}66 \times 10^6$ (liter/mole)^2/s^3, averaging $1{\cdot}55 \pm 0{\cdot}09$. The value of k_7 can be calculated since $k_1 = 8{\cdot}3 \times 10^{-6}\ s^{-1}$,[1] and $k_2 = 1{\cdot}0 \times 10^4$ liter/mole s.[8a] k_7 is found to be $1{\cdot}9 \pm 0{\cdot}1 \times 10^7$ liter/mole s.

Calculation of k_7 from the number of chain transfers per persulfate unimolecular decomposition

Another way of calculating k_7 is from the number of chain transfers per unimolecular persulfate decomposition. The number of chain transfers per

TABLE IV
Data and constants used for the calculation of chain transfers per radical

(I),[a] mole/liter H_2O	$2{\cdot}2 \times 10^{-4}$
Rp, mole/liter H_2O s	$5{\cdot}40 \times 10^{-4}$
(P), mole/liter H_2O	$5{\cdot}5 \times 10^{-7}$
D_a,[9,10] dm^2/s	2×10^{-7}
d, dm	$1{\cdot}17 \times 10^{-6}$
P_e [b]	0·82
V_{org}/V_{aq}	0·33

[a] $(I_0) = 3 \times 10^{-4}$ M, the value in this table is the initiator concentration at 50% conversion.
[b] P_e is the probability of escape $[= 1/(1 + k_4(M)ad^2/4D)]$ at 50% conversion.

radical can be calculated from the rate of polymerization. Table IV lists the data and constants used for this calculation. An (I_0) of 3×10^{-4} M was the mid-range of the concentrations used in measuring polymerization rates in Chapter 6, so it was used for the calculation. Particle concentration and volume of organic phase are also the intermediate values which were used in our kinetic study. The values for all parameters which may change as polymerization proceeds are chosen at 50% conversion.

$$Rp = k_2(P\cdot)(M) = 5{\cdot}40 \times 10^{-4} \text{ mole/liter s}$$

The polymerization rate constant was taken from the literature[8a] to be 10^4 liter/mole s; $[M]$ at 50% conversion is taken as 5 M. $(P\cdot)$ is then found to be $1{\cdot}08 \times 10^{-8}$ mole/liter H_2O.

The radical lifetime, τ, is calculated by dividing the average radical concentration by the rate of radical production (in this case, $2k_1[I] = 3{\cdot}7 \times 10^{-9}$ mole/liter s since $k_1 = 8{\cdot}3 \times 10^{-6}\ s^{-1}$)[1]

$$\tau = 1{\cdot}08 \times 10^{-8}/3{\cdot}7 \times 10^{-9} = 2{\cdot}9 \text{ s}$$

The chain transfer constant of vinyl acetate radical to monomer, k_3, is 2·0 liter/mole s.[8b] The number of chain transfers per radical is then $k_3[M]\tau = 2 \times 5 \times 2{\cdot}9 = 29$, We found that the rate of induced decomposition of initiator at 50% conversion was about four times higher than the unimolecular decomposition rate. Since each unimolecular decomposition produces two radicals, each radical generates an average of two induced decompositions before termination.

The rate of sweep-up of monomer radical by particles is $2\pi D_a d(P)$ (Chapter 6). The reciprocal of $2\pi D_a d(P)$ is then the time of sweep-up of monomer radical by particles. $1/(1-P_e)$ is the number of particles that monomer radical diffused through (P_e is the probability of escape of monomer radical from particle). The total time a given monomer radical spends in aqueous phase before termination is the number of particles that the monomer radical diffused through times the time of sweep-up of the monomer radical times the total number of chain transfers before termination, as is expressed in eqn. (5).

$$29\left(\frac{1}{1-P_e}\right)\left(\frac{1}{2\pi D_a d(P)}\right) = \text{total time of radical in aqueous phase} \quad (5)$$

The average time that monomer radical takes to react with initiator is given in eqn. (6)

$$\frac{1}{k_7(I)} = \text{average time for } (M\cdot_a) \text{ to react with initiator} \quad (6)$$

When (5) is divided by (6), the quotient should be equal to the number of induced decompositions per initial radical, eqn. (7).

$$\frac{29\,k_7(I)}{(1-P_e)\,2\pi D_a d(P)} = 2 \pm 0{\cdot}5 \quad (7)$$

All the parameters in eqn. (7) are known from Table IV except k_7, which is calculated to be $2{\cdot}8 \pm 0{\cdot}7 \times 10^7$ liter/mole s. Calculation of k_7 at other initiator concentrations (from $(I_0) = 1 \times 10^{-4}$ M to $(I_0) = 10 \times 10^{-4}$ M) gives values from $2{\cdot}6 \times 10^7$ to $3{\cdot}0 \times 10^7$ liter/mole s. The agreement of calculated values for k_7 is very good over the initiator range used. However, the rate of induced decomposition is small at 50% conversion and the error in the value used is relatively large.

Comparison of the four values for k_7

The results of calculation of the induced decomposition rate constant, k_7, using three independent methods (four ways of calculation) are shown in Table V. As can be seen, k_7 calculated from the rate of induced decomposition and chain transfer per persulfate initiation agree reasonably well. But the value of k_7 obtained from the correction term in the equation for the rate of polymerization is relatively high (Chapter 6). However, there is a large error (factor of three) associated with this determination. Within this broad error, k_7 agrees with the other values.

TABLE V
Comparison of k_7 values obtained by independent calculation methods

Method	k_7
1. Rate of induced decomposition	
a. (from K_2)	$2 \cdot 1 \pm 0 \cdot 05 \times 10^7$ liter/mole s
b. (from $fk_1k_2k_7$)	$1 \cdot 9 \pm 0 \cdot 1 \times 10^7$ liter/mole s
2. Rate of polymerization	2 to 12×10^7 liter/mole s
3. Chain transfer per persulfate initiation	$2 \cdot 8 \pm 0 \cdot 7 \times 10^7$ liter/mole s

The agreement between all the k_7 calculations is gratifying. Especially, the agreement between k_7s from 1a and 1b (Table V), plus the very small error range, show that the equations have the correct form and coefficients to describe initiator decomposition. This is strong evidence that the equations are correct theoretically, rather than simple curve fitting.

Proposed mechanism of induced decomposition

The consistency of k_7 shows that its high value must be accepted. The large rate constant indicates this reaction is probably diffusion controlled. It is therefore probable that the reaction goes by an electron transfer mechanism. In a recent paper, McGinniss and Kah[11] studied the interaction (electron transfer) of the $SO_4^{-}\cdot$ radical with different monomers. They showed that the greatest interaction of $SO_4^{-}\cdot$ was with monomers of low ionization potential (Ip) such as isobutyl vinyl ether ($Ip = 8 \cdot 74$ eV) and vinyl acetate ($Ip = 9 \cdot 19$ eV). The least interaction of $SO_4^{-}\cdot$ was with monomers of high ionization potential, such as methacrylonitrile ($Ip = 10 \cdot 37$ eV), methyl acrylate (10·72 eV), acrylonitrile (10·91 eV), etc.

Aromatic monomers having low Ip values (Ip ranges from 8·22 to 9·55 eV) give rate constants of about $10^9\,\mathrm{M}^{-1}\,\mathrm{s}^{-1}$ for interaction with $SO_4^{-}\cdot$. Aromatic compounds having Ip values of 9·7 to 9·92 eV give rate constants of 10^6 to $10^8\,\mathrm{M}^{-1}\,\mathrm{s}^{-1}$.[12]

The following equation was shown to be important for reactions of $SO_4^{-}\cdot$ and styrene ($Ip = 8 \cdot 47$ eV), hence this electron transfer mechanism from olefin to $SO_4^{-}\cdot$ is probably significant only with monomers having low Ip values.[13]

$$SO_4^{-}\cdot + \underset{\displaystyle R}{\underset{|}{CH}}{=}CH_2 \rightarrow SO_4^{-2} + \underset{\displaystyle R}{\underset{|}{\overset{(+)}{HC}}}{-}\dot{C}H_2 \qquad (8)$$

It is reasonable to propose an electron transfer mechanism in our case since the rate constant for the induced decomposition of initiator is comparable to the above cases. The proposed mechanism is shown below:

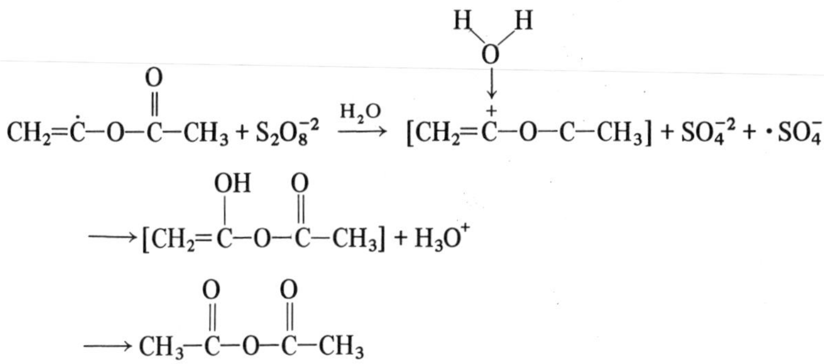

Monomer radical,

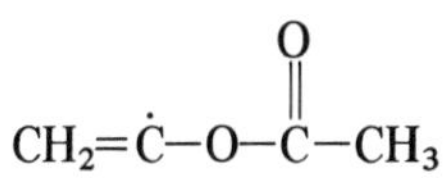

transfers an electron to $S_2O_8^{-2}$. A hydrated monomer cation is first formed which then hydrates to α-hydroxy vinyl acetate and finally rearranges to acetic anhydride or hydrolyses to two acetic acid molecules. This reaction starts with one stable high energy compound ($S_2O_8^{-2}$) and one unstable compound

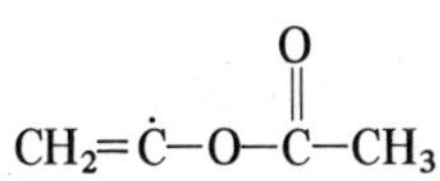

and generates two stable compounds (SO_4^{-2} and acetic anhydride) and one unstable compound ($\cdot SO_4^-$). Thus, overall the reaction should be exothermic. When the mechanism is compared to the reaction shown in eqn. (8), which starts from one stable

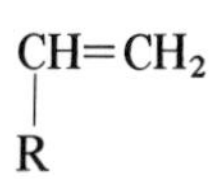

and one unstable compound ($SO_4^-\cdot$) and ends with one stable compound (SO_4^{-2}) and another unstable compound

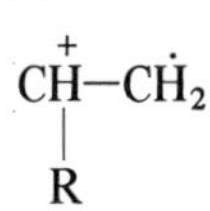

our proposed reaction probably proceeds more easily.

SUMMARY AND CONCLUSIONS

In this paper, we measured the initiator concentration as polymerization proceeded. The rate of induced decomposition of initiator was measured and found to depend on the number of particles to the 0·11 power and volume of the organic phase per unit volume of aqueous phase to the −0·64 power. The dependencies found agree very well with those predicted by our theory.

The rate constant for induced decomposition of initiator by the monomer radicals was found to be $2{\cdot}1 \pm 0{\cdot}1 \times 10^7$ liter/mole s. The high rate constant shows diffusion control of the reaction and implies that the reaction of persulfate with monomer radical in the aqueous phase proceeds through an electron transfer mechanism. Results from recent literature[10–12] show this mechanism to be reasonable.

In this chapter, we have shown that our proposed model for vinyl acetate emulsion polymerization predicts the right dependencies of induced decomposition of initiator on the various parameters. It also predicts the correct concentration/time curves. We showed in Chapter 6 that our proposed model predicted the dependence of polymerization rate on these parameters very well. Thus, there are now two lines of evidence, the dependencies of both the polymerization rate and rate of induced decomposition of initiator on various parameters, which agree with the theoretical predictions of the model. The wide range of predictive capability of the model supports the chemical steps proposed here.

APPENDIX A

From the kinetic model derived in Chapter 6, we have an expression for the concentration of monomer radical in the aqueous phase, $(M\cdot_a)$, eqn. (A1)

$$(M\cdot_a) = \frac{k_3 P_e(M)(2fk_1(I)(P))^{1/2}(k_4/k_3)^{1/2}(ad^2/4D)^{1/2}P_t^{-1/2}}{2\pi D_a d(P)\left[(1-P_e) + \dfrac{k_7(I)}{2\pi D_a d(P)}(1-2\bar{n})\right]} \times \left(1 + \frac{4Dk_7(I)}{k_4 2\pi D_a d^3(P)P_e(M)}\right)^{1/2} \tag{A1}$$

We have mentioned in the body of the text that on the average,

$$\frac{k_7(I)}{2\pi D_a d(P)}\bigg/(1-P_e) = 0{\cdot}07 \tag{A2}$$

Also from Chapter 6

$$\frac{P_e}{1-P_e} = \frac{4D/ad^2}{k_4(M)} \tag{A3}$$

$$V_{org}/V_{aq} = \frac{\pi d^3(P)N}{6} \tag{A4}$$

$$P_t^{-1/2} = \left(1 + \frac{2}{3}\frac{N\pi dD_0}{ak_{80}}\right)^{1/2} \tag{A5}$$

Substituting eqns. (A2), (A3), (A4) and (A5) into eqn. (A1), including explicitly the partition coefficient of the monomer radical between the organic and aqueous phase and grouping all constants, eqn. (A6) is obtained

$$\begin{aligned}(M\bullet_a) = {} & 1{\cdot}08 \times 10^{-9}(fk_1k_3/k_4)^{1/2}\left(\frac{D_0^{1/2}}{(M_0)}\right)\left(\frac{(M)^{3/4}}{D_a}\right) \\ & \times \left(1 + \frac{1{\cdot}422 \times 10^{23} D_0 d}{k_{80}(M)^{1/2}}\right)^{1/2} (P)^{1/6}\,(V_{org}/V_{aq})^{-2/3} \\ & \times (I)^{1/2}\left(1 + \frac{0{\cdot}072\,k_7 D_0(I)(M)^{1/2}}{k_4 D_a d^3(P)\,P_e(M_0)^2}\right)^{1/2}\end{aligned} \tag{A6}$$

We have shown in the previous chapter that

$$\left(1 + \frac{0{\cdot}072\,k_7 D_0(I)(M)^{1/2}}{k_4 D_a d^3(P)\,P_e(M_0)^2}\right)^{1/2}$$

contributes about 0·1 power dependence on (I), and thus

$$\frac{0{\cdot}072\,k_7 D_0(I)(M)^{1/2}}{k_4 D_a d^3(P)\,P_e(M_0)^2} \approx 0{\cdot}2$$

in the mid-range for (I). Thus the term

$$\left(1 + \frac{0{\cdot}072\,k_7 D_0(I)(M)^{1/2}}{k_4 D_a d^3(P)\,P_e(M_0)^2}\right)^{1/2} \approx 1{\cdot}1 \approx 1{\cdot}1\left(\frac{(I)}{(I_m)}\right)^{0{\cdot}1} \tag{A7}$$

Taking the mid-range value for (I), (I_m) at 50% conversion as $2{\cdot}2 \times 10^{-4}$ M and substituting eqn. (A7) into eqn. (A6), eqn. (A8) was obtained.

$$(M\cdot_a) \approx 2{\cdot}75 \times 10^{-9}(fk_1k_3/k_4)^{1/2}\,(D_0^{1/2}(M)^{3/4}/(M_0)\,D_a) \times \left(1 + \frac{1{\cdot}422 \times 10^{23} D_0 d}{k_{80}(M)^{1/2}}\right)^{1/2} (P)^{1/6}\,(V_{org}/V_{aq})^{-2/3}\,(I)^{0{\cdot}6} \tag{A8}$$

The change of initiator concentration with time is then the sum of unimolecular decomposition of initiator, $k_1(I)$, and the induced decomposition of initiator by the monomer radical in the aqueous phase.

$$\frac{-\,\mathrm{d}(I)}{\mathrm{d}t} = k_1(I) + k_7(M\cdot_a)(I) \tag{A9}$$

or

$$\frac{-\,\mathrm{d}\ln(I)}{\mathrm{d}t} = k_1 + k_7(M\cdot_a) \tag{A10}$$

Substituting for $(M\cdot_a)$ in eqn. (A10) gives

$$\frac{-\,\mathrm{d}\ln(I)}{\mathrm{d}t} = k_1 + 2{\cdot}75 \times 10^{-9} k_7(fk_1k_3/k_4)^{1/2}\,(D_0^{1/2}(M)^{3/4}/(M_0)\,D_a) \times \left(1 + \frac{1{\cdot}422 \times 10^{23} D_0 d}{k_{80}(M)^{1/2}}\right)^{1/2} (P)^{1/6}(V_{org}/V_{aq})^{-2/3}(I)^{0{\cdot}6} \tag{A11}$$

APPENDIX B

ANALYSIS OF THE INITIATOR CONCENTRATION AS A FUNCTION OF TIME

From zero time to 30% conversion

From Chapter 6 we have an expression for the rate of polymerization of vinyl acetate in emulsion:

$$Rp = 3{\cdot}09 \times 10^{-8} k_2\left(\frac{fk_1k_4}{k_3}\right)^{1/2} \left(1 + \frac{1{\cdot}422 \times 10^{23} D_0 d}{k_{80}(M)^{1/2}}\right)^{1/2} \times ((M_0)/D_0^{1/2})(M)^{1/4}\,(V_{org}/V_{aq})^{1/3}\,(P)^{1/6}\,(I)^{1/2}\left(1 + \frac{0{\cdot}072\,k_7D_0(I)(M)^{1/2}}{k_4D_ad^3(P)\;P_e(M_0)^2}\right)^{1/2} \tag{B1}$$

Below 30% conversion, every term in eqn. (B1) is constant except (V_{org}/V_{aq}) and the minor term,

$$\left(1+\frac{1\cdot422\times10^{23}D_0d}{k_{80}(M)^{1/2}}\right)^{1/2}$$

Because it is close to 1 and almost constant the average value for $(1+1\cdot422\times10^{23}D_0d/k_{80}(M)^{1/2})^{1/2}$ between 0 and 30% conversion can be used. Then we can write

$$Rp=-\frac{\mathrm{d}(M')}{\mathrm{d}t}=K_3(V_{org}/V_{orgf})^{1/3} \tag{B2}$$

where

$$K_3=3\cdot09\times10^{-8}\left(\frac{k_2(fk_1k_4)^{1/2}}{k_3^{1/2}}\right)((M_0)/D_0^{1/2})\left(1+\frac{1\cdot422\times10^{23}D_0d}{k_{80}(M_0)^{1/2}}\right)^{1/2}$$
$$\times(M_0)^{1/4}(P)^{1/6}(V_{orgf}/V_{aq})^{1/3}(I)^{1/2}\left(1+\frac{0\cdot072\,k_7D_0(I)(M_0)^{1/2}}{k_4D_ad^3(P)P_e(M_0)^2}\right)^{1/2} \tag{B3}$$

Since V_{org} equals V_{orgf} at 30% conversion, the following equation results

$$(V_{org}/V_{orgf})=\frac{c}{1-0\cdot7}=3\cdot33\,c=3\cdot33\left(\frac{(M_0')-(M')}{(M_0')}\right) \tag{B4}$$

where c = conversion of polymer, (M_0') = (total monomer + polymer in the seed)/liter H_2O, (M') = moles of monomer/liter H_2O during polymerization.

Substituting (B4) into (B2) and dividing both sides by (M_0'), (B5) is obtained.

$$\frac{\mathrm{d}((M')/(M_0'))}{\mathrm{d}t}=\frac{K_3}{(M_0')}(3\cdot33\,c)^{1/3} \tag{B5}$$

Also

$$\frac{-\mathrm{d}(M')}{(M_0')}=\frac{\mathrm{d}((M_0')-(M'))}{(M_0')}=\frac{\mathrm{d}c}{\mathrm{d}t} \tag{B6}$$

Substituting (B6) in (B5), (B7) is obtained.

$$\frac{\mathrm{d}c}{\mathrm{d}t}=\frac{K_3}{(M_0')}(3\cdot33)^{1/3}c^{1/3} \tag{B7}$$

Integrating (B7), (B8) is obtained

$$\tfrac{3}{2}(c^{2/3} - c_0^{2/3}) = \frac{K_3}{(M_0')}(3{\cdot}33)^{1/3}t \tag{B8}$$

Substituting (B4) into (B8) and rearranging the equation gives

$$\left(\frac{V_{\mathrm{org}}}{V_{\mathrm{orgf}}}\right)^{2/3} = \left(\frac{V_{\mathrm{org}_0}}{V_{\mathrm{orgf}}}\right)^{2/3} + \frac{2{\cdot}22\,K_3}{(M_0')}\,t \tag{B9}$$

V_{org_0} is the initial organic volume from the swollen polymer in the seed.

Since before 30% conversion, (M) and (P) are constant, and an average value for $(1 + 1{\cdot}422 \times 10^{23} D_0 d/k_{80}(M_0)^{1/2})^{1/2}$ is taken, (A11) (in Appendix A) can be written as follows

$$\frac{-\mathrm{d}\ln(I)}{\mathrm{d}t} = k_1 + 2{\cdot}75 \times 10^{-9} k_7 (fk_1k_3/k_4)^{1/2}\,(D_0^{1/2}(M_0)^{3/4}/(M_0)\,D_\mathrm{a})$$

$$\times\left(1 + \frac{1{\cdot}422 \times 10^{23} D_0 d}{k_{80}(M_0)^{1/2}}\right)^{1/2}(P)^{1/6}\,(I)^{0{\cdot}6}/(V_{\mathrm{org}}/V_{\mathrm{aq}})^{2/3} \tag{B10}$$

Substituting (B9) into (B10), (B11) is obtained.

$$\frac{-\mathrm{d}\ln(I)}{\mathrm{d}t} = k_1 + \left(\frac{2{\cdot}75 \times 10^{-9} k_7 (fk_1k_3/k_4)^{1/2}\,(D_0^{1/2}/(M_0))(1/D_\mathrm{a})(M_0)^{3/4}}{(V_{\mathrm{org}_0}/V_{\mathrm{orgf}})^{2/3}\,((M_0')/2{\cdot}22K_3) + t}\right)$$

$$\times\left(\frac{(1 + 1{\cdot}422 \times 10^{23} D_0 d/k_{80}(M_0)^{1/2})^{1/2}(P)^{1/6}((M_0')/2{\cdot}22K_3)(I)^{0{\cdot}6}}{(V_{\mathrm{orgf}}/V_{\mathrm{aq}})^{2/3}}\right) \tag{B11}$$

Equation (B11) can be numerically integrated.

The induced decomposition equation after 30% conversion

From (A11), the following equation results,

$$\frac{-\mathrm{d}\ln(I)}{\mathrm{d}t} = k_1 + K_2(D_0^{1/2}(M)^{3/4}/(M_0)\,D_\mathrm{a})\left(1 + \frac{1{\cdot}422 \times 10^{23} D_0 d}{k_{80}(M)^{1/2}}\right)^{1/2}$$

$$\times\,(P)^{1/6}\,(V_{\mathrm{org}}/V_{\mathrm{aq}})^{-2/3}\,(I)^{0{\cdot}6} \tag{B12}$$

The equation for the rate of polymerization has been given, (B1).

Integrating (B1) from (M_0) to (M) and time from t_0 to t, gives

$$(M_0)^{3/4} - (M)^{3/4} = 3{\cdot}09 \times 10^{-8} k_2 (fk_1 k_4/k_3)^{1/2} \, ((M_0)/D_0^{1/2}) \times \left(1 + \frac{1{\cdot}422 \times 10^{23} D_0 d}{k_{80}(M)^{1/2}}\right)^{1/2} \left(\frac{V_{\mathrm{org}}}{V_{\mathrm{aq}}}\right)^{1/3} (P)^{1/6} \times (I)^{1/2} \left(1 + \frac{0{\cdot}072\, k_7 D_0 (I)\, (M)^{1/2}}{k_4 D_a d^3 (P)\, P_e (M_0)^2}\right)^{1/2} (t - t_0) \qquad \text{(B13)}$$

Dividing (B13) on both sides by $(M_0)^{3/4}$, (B14) is obtained.

$$1 - ((M)/(M_0))^{3/4} = 3{\cdot}09 \times 10^{-8} k_2 (fk_1 k_4/k_3)^{1/2} \, ((M_0)^{1/4}/D_0^{1/2}) \times \left(1 + \frac{1{\cdot}422 \times 10^{23} D_0 d}{k_{80}(M)^{1/2}}\right)^{1/2} \left(\frac{V_{\mathrm{org}}}{V_{\mathrm{aq}}}\right)^{1/3} (P)^{1/6} \times (I)^{1/2} \left(1 + \frac{0{\cdot}072\, k_7 D_0 (I)(M)^{1/2}}{k_a D_a d^2 (P)\, P_e (M_0)^2}\right)^{1/2} (t - t_0) \qquad \text{(B14)}$$

Substituting (B14) into (B12), the final expression for $-\mathrm{d}\ln(I)/\mathrm{d}t$ is shown

$$\frac{-\mathrm{d}\ln(I)}{\mathrm{d}t} = k_1 + 2{\cdot}75 \times 10^{-9} k_7 (fk_1 k_3/k_4)^{1/2} \, (D_0^{1/2}/(M_0)^{1/4}) \left(\frac{1}{D_a}\right) \times \left(1 + \frac{1{\cdot}422 \times 10^{23} D_0 d}{k_{80}(M)^{1/2}}\right)^{1/2} (P)^{1/6} \, (V_{\mathrm{org}}/V_{\mathrm{aq}})^{-2/3} \, (I)^{0{\cdot}6} - 8{\cdot}48 \times 10^{-17} fk_1 k_2 k_7 \left(\frac{1}{D_a}\right) \left(1 + \frac{1{\cdot}422 \times 10^{23} D_0 d}{k_{80}(M)^{1/2}}\right)^{1/2} \times (V_{\mathrm{org}}/V_{\mathrm{aq}})^{-1/3} \, (P)^{1/3} \, (I)^{1{\cdot}2} \, (t - t_0) \qquad \text{(B15)}$$

(B15) can be numerically integrated and the coefficients of $(I)^{0{\cdot}6}$ and $(I)^{1{\cdot}2}$ found by minimization procedures.

REFERENCES

1. I. M. Kolthoff and I. K. Miller, *J. Amer. Chem. Soc.*, **73**, 3055 (1951).
2. W. V. Smith, *J. Amer. Chem. Soc.*, **71**, 4077 (1949).
3. W. V. Smith and H. N. Campbell, *J. Chem. Phys.*, **15**, 338 (1947).
4. I. M. Kolthoff, E. J. Meehan and C. I. Carr, *J. Amer. Chem. Soc.*, **75**, 1439 (1953).

5. P. D. Bartlett and K. J. Nozaki, *J. Polym. Sci.*, **3**, 216 (1948).
6. R. Patsiga, Thesis, New York State University, College of Forestry, Syracuse, New York, 1962.
7. K. H. S. Chang, M. H. Litt and A. M. Jamieson, The measurement of latex particle size by quasielastic laser light scattering, *J. Polym. Sci., A-2,* submitted for publication.
8. J. Brandrup and E. H. Immergut (eds.), *Polymer Handbook,* Interscience, New York, 1975, (a) p. II-63, (b) p. II-81.
9. M. Nomura, M. Harada, W. Eguchi and S. Nagata, ACS Symposium Series, No. 24, *Emulsion Polymerization* (1976).
10. C. R. Wilke and P. Chang, *AIChE Journal,* **1**, 264 (1955).
11. V. D. McGinniss and A. F. Kah, *J. Coatings Technology,* **49**, 61 (1977).
12. P. Neta, V. Madhavan, H. Zemel and R. W. Fessenden, *J. Amer. Chem. Soc.,* **99**, 163 (1977).
13. A. Ledwith and P. J. Russell, *Polym. Letts.,* **13**, 109 (1975).

8

The Reinvestigation of Vinyl Acetate Emulsion Polymerization (III) – Isotope Effect

M. H. LITT and K. H. S. CHANG

Department of Macromolecular Science, Case Western Reserve University, Ohio, USA

SUMMARY

To determine the major chain transfer site of vinyl acetate in emulsion polymerization, two deuterated vinyl acetates, vinyl trideuteroacetate and trideuterovinyl acetate were synthesized, polymerized both in bulk and emulsion and each compared with undeuterated vinyl acetate. The polymer molecular weights and polymerization rates of the three were determined under identical conditions. The major chain transfer site was found to be on the vinyl hydrogens of vinyl acetate (94%). The overall isotope effect of chain transfer to vinyl acetate was found to be 3·0. The major termination step in bulk polymerization was found to be the reaction of monomer radical with growing radicals.

INTRODUCTION

One of the most important factors in vinyl acetate polymerization is chain transfer of growing radical to the monomer.[1–5] This was concluded because there is no dependence of molecular weight on initiator concentration. While chain transfer to polymer was found, it is unimportant below 30% conversion; there is no molecular weight increase in the polymer until higher conversions. Hydrolysis and reacetylation of low conversion polymer shows no molecular weight change, indicating that there is no back-biting on polymer acetyl groups.

An interesting and to us an important question is where the main chain transfer site on vinyl acetate is. All studies concluded that the main chain transfer site was the hydrogen atoms of the acetyl group.[6–9] Some investigators found vinyl groups in the polymer and proposed some chain transfer on the vinyl hydrogens.[8,9] On the other hand, we concluded, on the basis of a kinetic analysis, that chain transfer is mainly on the vinyl hydrogens of vinyl acetate.[10]

One way to distinguish between the two possibilities is to study the isotope effect on the kinetics of vinyl acetate polymerization and on the polymer molecular weight. The deuterium isotope effect has been ascribed to the difference in the zero point energies of the stretching vibrations of the C–H and C–D bond.[11] The rate of a reaction in which deuterium is transferred is slower than that of the corresponding reaction for hydrogen, since the C–D bond has a lower zero point energy. The magnitude of the isotope effect is temperature dependent. For example, the reaction at a C–H bond has a maximum isotope effect of 7 at 25°C and 2·1 at 500°C.[12]

Urry[13] studied hydrogen abstraction from toluene α-d_1 versus toluene by chlorine, bromine and $CH_3COO\cdot$ radicals and found that the attack on toluene α-d_1 is slower than that on toluene. The isotope effect was found to be in the range of 2·2 to 9·9.

Bartlett and Tate[14] studied the polymerization of allyl acetate and allyl-1-d_2 acetate and found that the rate of polymerization for the deuterated allyl acetate was about 2·5 times as fast as that for the undeuterated allyl acetate. The molecular weight for the deuterated allyl acetate was 2·4 times as great as that of the undeuterated allyl acetate. An explanation was advanced that when deuterated allyl acetate was used there was less chain transfer on the monomer because of the stronger C–D bond. Both the polymerization rate and molecular weight increased by the same ratio when deuterated allyl acetate was used because in this sytem each radical produces one polymer molecule.

Thus, a relatively large isotope effect can be expected if deuterated vinyl acetate is used for polymerization. Since the degree of polymerization for vinyl acetate equals k_2/k_3, the ratio of the rate constant for propagation to that for chain transfer, the degree of polymerization should express the full isotope effect. It is known that in emulsion polymerization, $Rp \propto k_3^{-1/2}$,[10] and thus the relative change in rate should be proportional to the square root of the isotope effect. We felt that synthesis and polymerization of two partly deuterated vinyl acetates, trideuterovinyl acetate (D_2=CD—OAc) and vinyl trideuteroacetate

$$CH_2{=}CH{-}O{-}\overset{\overset{\displaystyle O}{\|}}{C}{-}CD_3$$

could settle the question of which end of the monomer is the major chain transfer site. If it is the acetyl group, when $CD_3COOCH=CH_2$ is used there will be an increase of the molecular weight compared to undeuterated vinyl acetate. On the other hand, if the polymerization of trideuterovinyl acetate shows the molecular weight increase, then the vinyl hydrogens are implicated. Also, it is expected that the emulsion polymerization rate would rise when trideuterovinyl acetate is used, if chain transfer on this site is as important as we believe.[10]

EXPERIMENTAL

Materials

Emulsifier

A very pure grade of sodium lauryl sulfate (Stepanol WA-100) was obtained from Stepan Chemical Company and was used directly from the bottle.

Initiator

Potassium persulfate was Fisher certified reagent grade and was used directly from the bottle in emulsion polymerization.

Seed latex

The seed latex was prepared as described before.[10] It had a solids content of 17·0% and its number average diameter was about 0·045 μm as measured by quasi-elastic laser light scattering.[15]

Vinyl acetate

Vinyl acetate was obtained from Eastman Kodak Company and purified by distillation through a Perkin Elmer spinning band column with 45 theoretical plates. The first and last 25% of the distillate was discarded for all the distillations. It was then stored at 5°C.

Vinyl trideuteroacetate ($CD_3{-}\overset{\overset{O}{\|}}{C}{-}O{-}CH{=}CH_2$)

The method of preparation was based on the interchange reaction of vinyl ester with acetic acid.[16,17] 150 g (1·5 mole) of vinyl propionate (Polyscience Inc.), 48 g (0·75 mole) of tetradeuteroacetic acid (Aldrich Chemical Co., 99·5 atom % D), 0·9 g of mercuric acetate, (Aldrich Chemical Co.), 0·155 g of concentrated sulfuric acid (Fisher Scientific, ACS reagent grade) and 0·001 g of hydroquinone (MC/B reagent grade) were put into a three-neck flask which was connected to a Perkin Elmer spinning band column with 45 theoretical plates. The mixture was heated to 90°C for two hours; 2 g of sodium acetate (Fisher Scientific, ACS reagent grade) was added and the resultant vinyl trideuteroacetate was distilled at atmospheric pressure. The fraction boiling from 71 to 75°C was collected. The crude vinyl trideuteroacetate was then redistilled and 58 g (0·65 moles) of pure vinyl acetate, bp 74°C, was obtained. The yield was 87% based on the tetradeuteroacetic acid used. The proton NMR spectrum of vinyl trideuteroacetate showed that the ratio of the proton peak intensities of methyl hydrogens to vinyl hydrogens was 0·018; the isotopic purity of

$$CD_3{-}\overset{\overset{\displaystyle O}{\|}}{C}{-}O{-}CH{=}CH_2$$

was 98·2%.

Trideuterovinyl acetate ($CH_3{-}\overset{\overset{O}{\|}}{C}{-}O{-}CD{=}CD_2$)

The reaction of acetylene with acetic acid[18, 19] was adopted for this synthesis. To a three-neck flask was added 30 g (1·50 moles) of deuterium oxide (Aldrich Chemical Co.), 162 g of acetic anhydride (1·59 moles) (Aldrich Chemical Co.) and 0·1 g conc. HCl (Fisher Scientific Co.). It was then stirred and heated to 100°C. After about 20 minutes, the reaction was complete; 180 g of monodeuteroacetic acid was formed. 7·0 g of mercuric acetate (Aldrich Chemical Co.) and 10 g of fuming sulfuric acid (Fisher Scientific Co.) were then added to the flask. Dideuteroacetylene was generated in another three-neck flask by the slow addition of 120 g of calcium carbide (Fisher Scientific, 80% pure) (1·15 moles) to 400 g of deuterium oxide and passed continuously for five hours at 20°C into the flask containing monodeuteroacetic acid and mercuric acetate. After five hours, 15 g of sodium acetate were added to the reaction mixture and the whole was distilled through a spinning band column at atmospheric pressure. Crude trideuterovinyl acetate was collected between 69 and 75°C. It was redistilled and 73 g of pure trideuterovinyl acetate, bp 73°C, was obtained. The yield was 55% based on the dideuteroacetylene used. The proton NMR spectrum of trideuterovinyl acetate showed that the ratio of proton peak intensities of vinyl to methyl hydrogens was 0·034; the isotopic purity of

$$CH_3{-}\overset{\overset{\displaystyle O}{\|}}{C}{-}O{-}CD{=}CD_2$$

was 96·6%.

Polymerizations

Bulk polymerization

Freshly distilled monomer, 3·5 g, was put into a 12 ml polymerization tube which was then degassed under a vacuum of 10^{-3} mm Hg four or five times and sealed. Three sealed samples of each monomer were immersed in 700 g of molten $NaH_2PO_4 \cdot 2H_2O$, mp = 60·0°C (made by adding 80·78 g of H_2O to 619·22 g of $NaH_2PO_4 \cdot H_2O$) at 60°C in a well-insulated beaker. The tubes were then irradiated at $0{\cdot}12 \times 10^6$ rad/h using a ^{60}Co γ-ray source and samples were withdrawn at selected intervals. Conversion was measured by evaporating the monomer at 1 mm Hg and 40°C for 8 h, raising the temperature to 65°C and weighing the residue at intervals until it reached constant weight.

Emulsion polymerization

The polymerization apparatus was a dilatometer which was similar to, though smaller than, the one used before.[10] A graduated 10 ml column (adapted from a 10 ml pipette) was connected to a 50 ml volumetric flask with three indentations in it set at 120°. Latex volume changes were read by taking the emulsion to a reference point in the column and reading the height on a precision bore side tube.

The seeded polymerization procedure was the same as described earlier[10] except that only 0·175 of the recipe was used. Polymerizations of all monomers were run at 60°C using four different initial initiator concentrations ($(I_0) = 2 \times 10^{-4}$ M to 10×10^{-4} M) at a constant volume of organic phase per unit volume of aqueous phase ($V_{org}/V_{aq} = 0{\cdot}33$). Rates were calculated from the slopes of the linear portion of the polymer/time curves at 30–85% conversion.

Analysis

NMR measurement

Monomer, 0·3 g, was dissolved in 1 g of $CDCl_3$ using TMS as internal standard. NMR spectra were run using a Varian A-60A proton NMR instrument at 23°C.

Viscometry

The specific viscosity of each polymer from the bulk polymerization was measured in acetone at 30°C using an Ubbelohde dilution viscometer. Five concentrations in the range of 1·120 to 0·242 g/dl poly(vinyl acetate) and poly(vinyl trideuteroacetate) and 0·385 to 0·084 g/dl (poly(trideuterovinyl acetate)) were run. Intrinsic viscosity was calculated by extrapolation of the η_{sp}/c versus c plot to zero concentration. Molecular weights were calculated using the equation,[20] $[\eta] = 1 \times 10^{-4} [M_n]^{0.72}$, which is in about the middle of the range of ks and α found.

RESULTS AND DISCUSSION

Polymerization rate

Bulk polymerization

The conversion versus time plots for the bulk polymerizations of vinyl acetate and its deuterated analogues are shown in Fig. 1. Vinyl trideuteroacetate has a conversion rate of $9{\cdot}9 \times 10^{-3}$/min which is identical with that of vinyl acetate ($9{\cdot}5 \times 10^{-3}$/min) within the experimental error. However, trideuterovinyl acetate has a much higher conversion rate ($1{\cdot}69 \times 10^{-2}$/min).

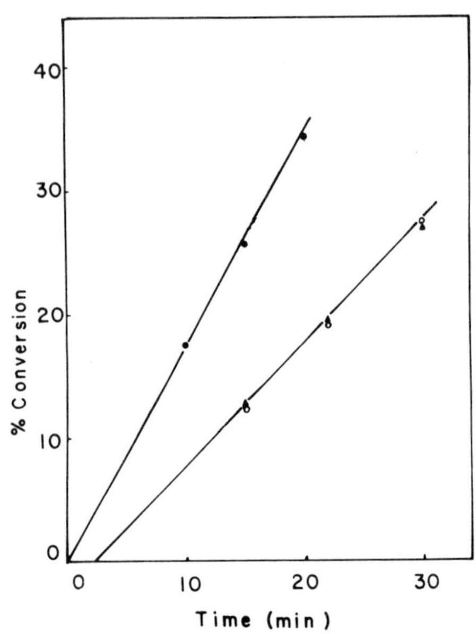

Fig. 1. The plot of % conversion versus time in bulk polymerization for three monomers:

●, $CD_2{=}CD{-}O{-}C({=}O){-}CH_3$; ▲, $CH_3{-}C({=}O){-}O{-}CH{=}CH_2$; ○, $CD_3{-}C({=}O){-}O{-}CH{=}CH_2$.

The ratio of the rate of polymerization of trideuterovinyl acetate to the average of the other two monomers is 1·74 ± 0·03.

Emulsion polymerization

Table I shows the comparison of rates of emulsion polymerization of vinyl acetate and its deuterated analogues as the initial initiator concentration, (I_0), was varied. Again, the rate of polymerization for vinyl trideuteroacetate is identical with that of vinyl acetate, while the rate of polymerization of trideuterovinyl acetate averages 1·76 times higher. This indicates that vinyl

TABLE I
The comparison of seeded emulsion polymerization, Rp, of three monomers (vinyl acetate, vinyl trideuteroacetate, and trideuterovinyl acetate) at various initiator concentrations

	$10^4(I_0)$ $(m/liter)$			
Monomer/100 Rp (M/*min*)	2·0	3·0	5·0	10.0
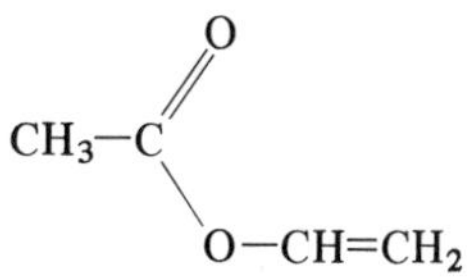	2·63	3·28	4·44	6·89
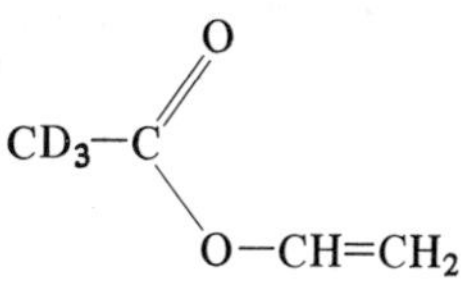	2·59	3·26	4·54	6·79
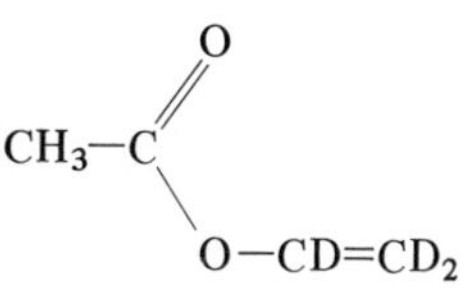	4·57	5·85	7·77	12·27
Rp_3/Rp_1[a]	1·74	1·78	1·75	1·78

[a] Rp_3 and Rp_1 are the polymerization rates of 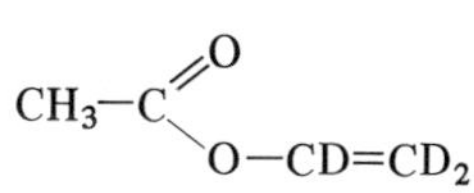 and 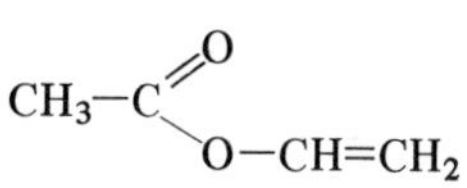 respectively.

hydrogen is the major chain transfer site in vinyl acetate. Also, Table I shows that the chain transfer on acetyl hydrogens (shown later to be 6%) is kinetically insignificant. The accuracy of measurement of the rate is about 1% and the ratio of the rates of polymerization of vinyl acetate and vinyl trideuteroacetate average 1·00 to within $\frac{1}{2}$%. If the chain transfer step is kinetically significant, the rate should have been increased by 2%, which would be easily seen. Our previous work[10, 22] showed this step should have no kinetic effect.

Molecular weights

The molecular weights of the three polymers at three different conversions are compared in Table II. As can be seen, the molecular weight of poly(trideuterovinyl acetate) averages 2·59 times higher than that of poly(vinyl acetate) indicating again that the major chain transfer site is the vinyl

TABLE II

The intrinsic viscosities and molecular weights of poly(vinyl acetate), poly(vinyl trideuteroacetate), and poly(trideuterovinyl acetate) at three conversions

Polymer	*Conversion*	$[\eta]$ (*dl/g*)	*Molecular weight*[a]	M_W/M_{W_1} [b]
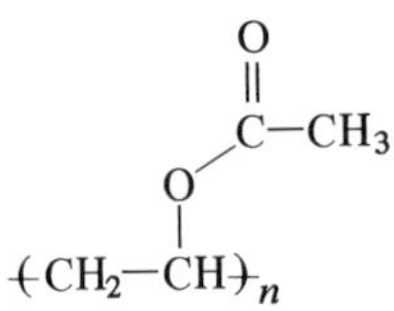	0·127	1·15	$4 \cdot 37 \times 10^5$	—
	0·195	1·17	$4 \cdot 48 \times 10^5$	—
	0·274	1·16	$4 \cdot 42 \times 10^5$	—
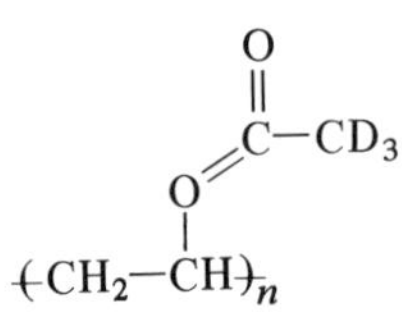	0·125	1·18	$4 \cdot 57 \times 10^5$	1·05
	0·193	1·19	$4 \cdot 61 \times 10^5$	1·03
	0·274	1·19	$4 \cdot 60 \times 10^5$	1·04
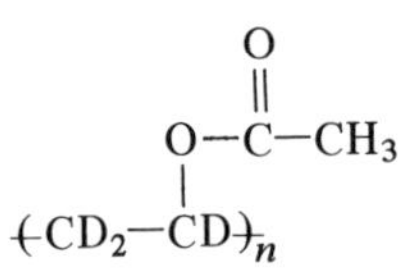	0·175	2·28	$11 \cdot 31 \times 10^5$	2·59
	0·257	2·32	$11 \cdot 58 \times 10^5$	2·58
	0·344	2·31	$11 \cdot 51 \times 10^5$	2·60

[a] Molecular weight was calculated by $[\eta] = KM^a$, where $K = 1 \cdot 0 \times 10^{-4}$ (dl/g), $a = 0 \cdot 72$.[20]

[b] M_W/M_{W_1} is the ratio of molecular weight of deuterated vinyl acetate to that of vinyl acetate.

hydrogens. Also there is some chain transfer on the acetyl hydrogens of vinyl acetate since the molecular weight of poly(vinyl trideuteroacetate) is about 1·04 times that of poly(vinyl acetate).

The fraction of chain transfer on vinyl hydrogens and the magnitude of the isotope effect

If chain transfer is on both vinyl and acetyl hydrogen, the degree of polymerization of the polymers can be expressed as in eqn. (1)

$$\bar{X}_n = \frac{k_2}{k_{3A} + k_{3V}} = \frac{1}{C_{MA} + C_{MV}} \quad (1)$$

where C_{MA} and C_{MV} are the chain transfer constants on acetyl and vinyl hydrogens of vinyl acetate respectively. The kinetic model is given in the next section.

$$C_{MA} = \frac{k_{3A}}{k_2} \quad (2)$$

$$C_{MV} = \frac{k_{3V}}{k_2} \quad (3)$$

We know from Table II that the molecular weight of poly(vinyl trideuteroacetate) is 1·04 times that of poly(vinyl acetate).

Thus, taking into account isotopic purity,

$$\frac{\bar{X}_{nDA}}{\bar{X}_n} = 1{\cdot}04 = \frac{C_{MA} + C_{MV}}{0{\cdot}982C_{MA}/x + 0{\cdot}018C_{MA} + C_{MV}} \quad (4)$$

where $\bar{X}_n$ and $\bar{X}_{nDA}$ are the degrees of polymerization of vinyl acetate and vinyl trideuteroacetate respectively and x is the isotope effect.

Also, the molecular weight of poly(vinyl trideuteroacetate) is 2·59 times that of poly(vinyl acetate). Taking into account the remaining hydrogen, eqn. (5) results

$$\frac{\bar{X}_{nDV}}{\bar{X}_n} = 2{\cdot}59 = \frac{C_{MA} + C_{MV}}{C_{MA} + 0{\cdot}966C_{MV}/x + 0{\cdot}034C_{MV}} \quad (5)$$

where $\bar{X}_{nDV}$ is the degree of polymerization of trideuterovinyl acetate.

Solving (4) and (5) for the unknowns we obtain

$$\frac{C_{MA}}{C_{MV}} = 0{\cdot}062 \quad (6)$$

$$x = 3{\cdot}04 \quad (7)$$

Thus, only 6% of the chain transfer is on the acetyl hydrogens and 94% on the vinyl hydrogens of vinyl acetate. The overall isotope effect is 3·04.

The rate of polymerization in emulsion polymerization is proportional to $k_3^{-1/2}$, where k_3 is the chain transfer step on the vinyl group.[10] Substituting trideuterovinyl acetate for vinyl acetate raised the rate by a factor of 1·76. When the calculation for the isotope effect on rate is done accurately, taking into account the 3% H on the trideuterovinyl, we find that if the effect is purely on k_3, the rate should rise by a factor of 1·69 as compared to 1·76 ± 0·02. This is almost within the experimental error. There may be a very slight secondary isotope effect[23,24] on the propagation and reinitiation rate constants k_2 and k_4, but it cannot be decided from these data.

Model and nature of termination step for bulk polymerization of vinyl acetate

The following is a reasonable kinetic model for the bulk polymerization of vinyl acetate.

1. $I \xrightarrow{k_1} R\cdot$
2. $R_n^{\cdot} + M \xrightarrow{k_2} R_{n+1}^{\cdot}$ — $R\cdot \equiv$ Growing radical
3. $R\cdot + M \xrightarrow{k_3} RH + M\cdot$ — $M\cdot \equiv$ Monomer radical from chain transfer
4. $M\cdot + M \rightarrow R\cdot$
5. $R\cdot + R\cdot \rightarrow x$ (termination)
6. $R\cdot + M\cdot \rightarrow x$ (termination)

If k_4 is relatively large, e.g. 10^3, then $M\cdot$ would be present in such low concentration that step 5 would be the major termination step. If k_4 is *very* small (little or no reinitiation) then step 6 plus $M\cdot + M\cdot$ would be the major termination steps. Such a case leads to allyl acetate type of kinetics with both $\bar{x}$ and Rp proportional to $1/k_3$,[14] which is not found. Thus $M\cdot + M\cdot$ cannot be considered as a major termination step. We shall solve for extremes of the remaining termination steps which imply a reasonable amount of reinitiation, steps 5 and 6, assuming that each separately occurs almost 100%.

(A) The expressions for the rate and degree of polymerization when the major termination step is step 5 are given in eqns. (8) and (9), respectively. The solution is based on the steady state assumption, $k_3(R\cdot)(M) = k_4(M\cdot)(M)$, which implies that the rates of steps 3 and 4 are at least five times faster than termination.

$$Rp = k_2 \left(\frac{k_1}{2k_5}\right)^{1/2} (I)^{1/2} (M) \tag{8}$$

$$\bar{x}_n \approx \frac{k_2}{k_3} \tag{9}$$

In this case, deuteration has little effect on the polymerization rate since k_3 (chain transfer constant) does not appear in eqn. (8). But there will be an increase in molecular weight compared to vinyl acetate since the chain transfer constant for trideuterovinyl acetate is smaller than that for vinyl acetate.

(B) The major termination is step 6; the rate of polymerization is now given by (10)

$$Rp = k_2 \left(\frac{k_1 k_4}{2k_3 k_6}\right)^{1/2} (I)^{1/2} (M) \tag{10}$$

The equation for degree of polymerization is unchanged, (9). Thus an increase of polymerization rates by $(k_3/k_{3D})^{1/2}$ and molecular weight by k_3/k_{3D} should be observed when trideuterovinyl acetate is used (k_{3D} is the chain transfer rate constant for trideuterovinyl acetate).

We have seen in Tables I and II that both the polymerization rate and molecular weight of trideuterovinyl acetate increased compared to that of vinyl acetate. Since the rate increase is identical for bulk and emulsion polymerization, and we know that almost all termination in emulsion polymerization is between monomer and growing radical,[10] the implication is that the same is true in bulk polymerization. While this conclusion is initially surprising, it supports our argument that the vinyl radical is reasonably stable.[10] Otherwise the radical would reinitiate rapidly and its concentration would be too low for termination with it to be important.

Further consideration shows that one should expect almost all termination to be by step 6. The steady state assumption can be written as

$$(R\cdot)/(M\cdot) = k_4/k_3 \tag{11}$$

k_4 was found[10] to be 29 while k_3 is known from the literature to be 2 to 2·5 at 60°C (when $k_2 \approx 10^4$). Thus, $(R\cdot)/(M\cdot) \approx 13 \pm 2$. Since $R\cdot$ is a macromolecule with average degree of polymerization of four to five thousand while $M\cdot$ is a monomer radical, $M\cdot$ has a diffusion constant at least 60 to 100 times greater than $R\cdot$. Since termination rate constants are diffusion controlled, k_5 will be much smaller than k_6. Even though $R\cdot$ is the majority radical present, cross-termination is favored; $k_5(R\cdot)^2/k_6(M\cdot)(R\cdot) \leqslant 1/5$. On the other hand, the termination of two $M\cdot$s is disfavored even though this termination rate constant should be twice as high as k_6, since both radicals are small. Here we find $k_6(M\cdot)(R\cdot)/2k_6(M\cdot)^2 = (R\cdot)/2(M\cdot) \approx 6$. Thus by chance, we are in a region where even with no cross-termination enhancement (which usually occurs), the major termination must be by step 6, considering the rate constants found in our previous paper to be correct.

SUMMARY AND CONCLUSIONS

The bulk polymerization rate of trideuterovinyl acetate was found to be 1·78 times that of vinyl acetate. In emulsion polymerization the same result was found. The molecular weight of poly(trideuterovinyl acetate) was 2·59 times that of poly(vinyl acetate). The overall isotope effect on chain transfer to monomer was calculated to be 3·04. Chain transfer was shown to be 94% on the vinyl hydrogens and 6% on the acetyl hydrogens. The measurement of polymerization rates in emulsion polymerization showed that the chain transfer on acetyl hydrogens is kinetically insignificant.

The results definitely prove our hypotheses in the kinetic model for vinyl acetate emulsion polymerization,[10] that vinyl radical, $CH_2{=}\dot{C}{-}OAc$, is the major monomer radical formed and is a stable radical which reinitiates relatively slowly compared to the propagation step.

The idea that 94% of the chain transfer at 60°C is on vinyl hydrogens is completely different from the conclusions of all previous workers. However, they had to use indirect methods to determine the chain transfer site. Here deuterated vinyl acetates tested the isotope effect on the potential chain transfer sites directly and gave an unequivocal answer.

The termination step in bulk polymerization, which was previously assumed to be growing radical reacting with growing radical, has been shown here to be almost completely monomer radical reacting with growing radical.

ACKNOWLEDGEMENTS

The authors thank the Polymer Section of the Materials Science Division for partial support of this work on Grant No. DMR76-04562. We particularly want to thank Dr W. E. Daniels of Air Products Corporation for suggesting the use of deuterated monomers.

REFERENCES

1. O. L. Wheeler, S. L. Ernst and R. N. Crozier, *J. Polym. Sci.,* **VIII**, 409 (1952).
2. M. Matsumoto, K. Imai, M. Maeda, Y. Ohyanagi and T. Saito, *Kobunshi Kagaku,* **12**, 398 (1955).
3. K. Imai, *J. Chem. Soc. Japan, Ind. Chem. Sect.,* **62**, 1127 (1959).
4. M. Matsumoto and Y. Ohyanagi, *J. Polym. Sci.,* **XLVI**, issue 148 (1960).
5. (a) N. Friis and L. J. Nyhagen, *J. Appl. Polym. Sci.,* **17**, 2311 (1973).
   (b) N. Friis, Thesis, Technical University, Copenhagen, Denmark, 1973.
6. M. Lazar, J. Paulinec and Z. Manasek, *Collection Czech. Chem. Commun.,* **26**, 1380 (1961).
7. S. Imoto, J. Ukida and T. Kominami, *Kobunshi. Kagaku,* **14**, 101 (1957).
8. O. L. Wheeler, E. Lavin and R. N. Crozier, *J. Polym. Sci.,* **9**, 157 (1952).
9. R. O. Howard, Ph.D. Thesis, Massachusetts Institute of Technology, Cambridge, Mass., 1952.
10. K. H. S. Chang, M. H. Litt and M. Nomura, this volume, Chapt. 6.
11. W. P. Jenks, *Catalysis in Chemistry and Enzymology*, McGraw-Hill Book Co., New York, 1969.
12. K. B. Wiberg, *Chem. Rev.,* **55**, 713 (1955).
13. W. H. Urry, *Abstracts 12th National Organic Symposium, Denver, Colorado, June, 1951,* p. 30.
14. P. D. Bartlett and F. A. Tate, *J. Amer. Chem. Soc.,* **75**, 91 (1953).
15. K. H. S. Chang, M. H. Litt and A. M. Jamieson, The measurement of latex particle size by quasielastic laser light scattering, *J. Polym. Sci., A2,* submitted for publication.
16. W. Toussaint and G. MacDowell, US Patent 2,299,862.
17. R. Adelman, *J. Org. Chem.,* **14**, 1057 (1949).
18. H. B. Dykstra, US Patent 1,849,616.
19. F. W. Skirrow and G. O. Morrison, US Patent 1,855,366.
20. J. Brandrup and E. H. Immergut (eds.), *Polymer Handbook*, Interscience, New York, 1975, p. IV-15.
21. W. J. Jenks, *Catalysis in Chemistry and Enzymology*, McGraw-Hill Book Co., New York, 1969, p. 253.
22. M. H. Litt, R. Patsiga and V. T. Stannett, *J. Polym. Sci., A1,* **8**, 3607 (1970).
23. E. R. Thornton, *Ann. Rev. Phys. Chem.,* **17**, 349 (1966).
24. A. Streitwieser, Jr., R. H. Jagow, R. C. Fahey and S. Suzuki, *J. Amer. Chem. Soc.,* **80**, 2326 (1958).

9

The Radiation Induced Emulsion Polymerization of Vinyl Acetate

V. T. STANNETT, R. R. CHALLA, J. H. DREW and E. P. STAHEL
Department of Chemical Engineering, North Carolina State University, USA

SUMMARY

The advantages and disadvantages of using high energy radiation to initiate the emulsion polymerization of vinyl monomers in general, and vinyl acetate in particular, are discussed. Since the activation energy for the initiation step is close to zero, polymerizations can be conducted at any temperature without resorting to complex redox systems. In the case of vinyl acetate, low temperature polymerizations lead to high molecular weights with considerable advantages for some paint and coating applications.

The kinetic and other features of the polymerization are discussed in some detail. Descriptions are given of a number of pilot plant studies using flow systems and some of the results obtained. The properties of such latexes are presented and compared with a corresponding commercial product made with chemical initiation.

INTRODUCTION

The polymerization of vinyl monomers in emulsion has been practised with many monomers and monomer mixtures over many years on a large industrial scale. Vinyl acetate and its copolymers have been prominent among these large volume productions. All of these polymerizations are free radical in nature and involve the use of chemical initiators, normally potassium persulfate or various redox systems. The use of high energy radiation has been studied comparatively little compared with the huge volume of literature on chemically initiated systems, and does not appear to be practised at all on an industrial scale.

There are a number of inherent advantages associated with the use of high energy radiation for the initiation of vinyl monomer polymerization in emulsion compared with chemical methods. Vinyl acetate may have particular interest in these respects. This paper will discuss the radiation induced polymerization of vinyl acetate in emulsion both from the point of view of basic laboratory investigations and from a number of pilot plant studies.

In general, Cobalt-60 gamma radiation has been used because of its high degree of penetration, comparative ease of estimating the dose-depth characteristics and because radical fluxes comparable to those used with chemical initiation can easily be achieved. There have also been a few comparatively brief studies using electron accelerators to initiate emulsion polymerization in emulsion.

It is worthwhile at this stage to discuss some of the reasons why the authors feel that radiation initiation does have a number of advantages over chemical initiation, particularly with vinyl acetate, but also in general.

The yield of free radicals from the radiolysis of water is high, of the order of six radicals per 100 eV of absorbed energy. This compares, for example, with about 0·7 for bulk styrene. Combined with the high kinetic chain lengths associated with emulsion polymerizations in general, this leads to a cost of initiation at least comparable with chemical systems. Radiation, particularly using isotopes such as Cobalt-60, can give an essentially unlimited range of radical fluxes, from zero to those equivalent to many moles per liter of chemical initiators, which would clearly be impractical. Associated with this range is the ease with which the fluxes can be monitored during the course of the polymerization reaction as the kinetics and other considerations demand. The fluxes can be programmed and built into modern process control techniques leading in principle to the orderly control of molecular weight and partical size distributions and the elimination of residual monomer. There need be no change in the radical flux during the reaction, i.e. no initiator exhaustion and the initiation process can be started and stopped at will by simply dropping away or raising the radiation source. This could be useful for changes in the feed in copolymerizations and, in principle, for block and graft polymerization in emulsion.

Initiation with radiation has no temperature dependence. This leads to comparatively low temperature dependencies for the overall reaction, the activation energy dropping from about 20 for chemical initiation to only, say, 7 kcal/mole. Such a change, especially coupled with the ease of removing the initiation source, makes the possibilities of exothermic, runaway reactions extremely low indeed. Furthermore, initiation is not a function of polymerization temperature and hence is completely uncoupled, in direct contrast to chemical initiation.

The radicals produced by the radiolysis of water are hydrogen atoms and, mainly, hydroxyl radicals. These are neutral and highly reactive radicals leading to highly efficient initiation, there are no electrolytes such as arise from most chemical systems, this leads to higher surface tension latexes with lower ionic strengths and higher pHs, often eliminating the need for adding buffers as is necessary with potassium persulfate initiation, for example. No contamination with residual initiator fragments occurs. Another advantage which could conceivably become of considerable importance in the future is that radiation is ideal in initiating the polymerization of systems using cationic soaps which often interact with chemical systems. This feature was pointed out and investigated many years ago in Japan by Okamura and his co-workers.

The lack of any activation energy with the initiation reaction has already been emphasized. This also means that polymerizations can be conducted at will at any temperature and at any initiation rate. In principle, this can be, and indeed is accomplished with chemical initiation including redox systems. In practice, however, especially low temperature initiation is not easy to achieve and control, particularly with polar monomers such as vinyl acetate. The low temperature polymerization of vinyl acetate is of particular interest. It is well known that monomers such as vinyl acetate and vinyl chloride, whose degree of polymerization is mainly governed by chain transfer to monomer, have a negative dependence of the molecular weight on temperature. To achieve high molecular weights therefore, low temperature polymerizations are necessary.

Poly(vinyl acetate) latexes of high molecular weight have a number of industrially interesting advantages and these could be more comfortably produced by radiation. This has led over the years to a considerable interest in its production using radiation.

There are disadvantages attached to the use of radiation; there are, for example, no ionic end groups such as arise from persulfate initiation. These could, in principle, lead to some stabilization of the resulting latex. In addition, the radiation attacks all the components including the emulsifier and the polymer as it is formed. The former could lead to a very small amount of grafting but is probably a negligible factor. The latter, however, leads to branching plus a very small amount of lost acetate groups. It is doubtful, however, that under the conditions which would prevail in an industrial process any of these problems could cause difficulties. A more important problem, which could arise in flow reactors, or even kettle systems, is possible build up of polymer on the walls of the reaction vessel due to the diminishing flow rate approaching the walls. This could also arise since the radiation is at its most intense near the radiation elements. It is felt, however, that with suitable reactor designs this problem could also be eliminated, as with general coating situations encountered with chemical initiation. In fact with kettle reactors these effects have not been observed in practice, at least over the limited time periods involved in pilot plant studies. With wide tube flow reactors some coating of the walls has been observed but again it is probably not a major problem, at least over reasonable time periods.

LABORATORY STUDIES

The earliest studies of the gamma radiation polymerization of vinyl acetate in emulsion appear to be those of Allen *et al.*[1] Closely followed by Okamura *et al.*,[2] Allen *et al.* prepared poly(vinyl acetate) latex using gamma radiation in order to use the trapped radicals remaining after the source was removed to generate block copolymers by the subsequent addition of methyl methacrylate. A considerable amount of block copolymer was, indeed, formed. A G (-monomer) value, i.e. the number of monomer units polymerized per 100 eV of absorbed radiation, of about 10^5 was found with the emulsion polymerization itself at a dose rate of 0·06 Mrad/h. In more detailed accounts[3] the conversion curves were found to be linear for about 15–70% polymerization and the molecular weights were of the order of 10^6. The specific rate constants for chain propagation, based on the assumption that the simple Smith–Ewart Case II kinetics were applicable were estimated. The values found were only about $\frac{1}{4}$ to $\frac{1}{2}$ of those determined by other techniques. It was correctly assumed that less than 0·5 radicals per particle were present, and the mechanism was closer to Smith-Ewart Case I.

The work of Okamura *et al.*[2] concentrated more on the effect of different emulsifiers, those studied were anionic (sodium lauryl benzene sulfonate), nonionic (polyoxyethylene lauryl ether) and cationic (cetyl pyridinium chloride and stearyl ammonium chloride). The rates of polymerization at the same emulsifier concentration and dose rate were found to be slower in the case of the cationic emulsions, the pyridinium salt being by far the lowest. There followed reports of extensive studies of similar systems.[4,5]

The principal additional findings were that the activation energy for the rates of polymerization was 5·7 kcal/mole, the rates increased with increasing dose rate but with an unspecified dependency. The dependence of the rate and the molecular weights on the emulsifier concentrations was found to be between 0·14 and 0·40 order. Stable cationic latexes up to 50% by weight were obtained. A more thorough investigation of the stability of radiation initiated poly(vinyl acetate) latexes has also been reported by Okamura and co-workers.[6,7]

Finally a series of studies has been reported in which vinyl acetate was polymerized in emulsion using an electron accelerator.[8-11] The polymerizations tended to level off at about 90% conversion. Particle sizes were extremely small and at 20% solids the latexes were almost transparent. The degrees of polymerization were about 2500 and essentially independent of the dose rate. Continuous and pulsed radiation were investigated, the latter gave several times faster rates.

RATE STUDIES

The first systematic study of the kinetics of vinyl acetate polymerization in emulsion initiated by gamma radiation was reported by Stannett *et al.*[12] in 1967. Most of the investigation was conducted using sodium lauryl sulfate as the emulsifying agent. Careful studies of this system had previously been carried out by the same author and his colleagues using potassium persulfate as the initiator.

Variables studied included temperature, water to monomer ratio, emulsifier concentration and dose rate. The polymerization rates, molecular weights and the number of particles were measured. A few seeded polymerizations were also conducted. Conversions were very smooth and proceeded to more than 99% conversion. Typical conversion curves at 30°C for both vinyl acetate and styrene are presented in Fig. 1. Both were very rapid. The vinyl acetate curve was essentially linear up to about 85% conversion whereas styrene showed the well-known slightly S-shaped curve. These differences have also been found with potassium persulfate initiated polymerizations of these two monomers and have been discussed at length in the literature. In addition, vinyl acetate, although polymerizing more rapidly than styrene was not as fast compared with styrene as would be predicted by their different absolute rate constants for propagation. This had previously been commented on by Allen *et al.*[1,3] and by Hummel[13] and ascribed to the different mechanisms by which the polymerizations proceed.

A number of experiments were conducted up to 50% conversion in which the irradiation source was removed for 30 minutes and then returned. The polymerizations stopped almost immediately and then continued to proceed at the normal rate on restoration of the radiation source. Similar results were found by Friis[14] on his flow, pilot plant system; more recently Sunardi[15] has studied the gamma radiation initiated emulsion polymerization of vinyl acetate under intermittent conditions. It was found that with five minutes without irradiation no breaks in the conversion curves occurred, but with

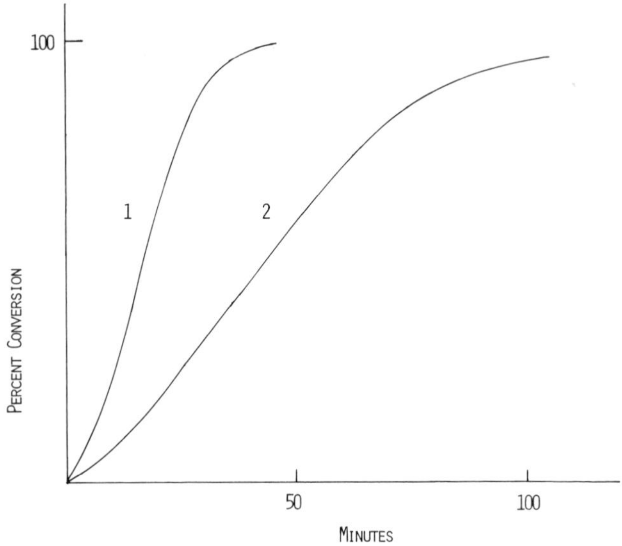

Fig. 1. Conversion–time curves for the radiation induced emulsion polymerization of vinyl acetate and styrene at 30°C. Dose rate 0·02 Mrad/h. 1, Vinyl acetate; 2, styrene.[12]

15 minutes, step-like curves were found. This would be consistent with the results of Stannett *et al.*[12] where 30 minute intervals without radiation were employed. The results of Friis are also consistent since a very small increase in conversion was observed followed by a cessation of polymerization. Allen *et al.*[1,3] had demonstrated the presence of trapped radicals by grafting studies at 70% conversion and above. Undoubtedly at the lower conversions the radicals chain transferred to monomer, a very rapid process; and in the more highly swollen polymer, particles soon transferred out into the aqueous phase and terminated, essentially stopping the polymerizations. The consequences of 'trapped' radicals can be considerable, however, particularly with flow reactors as described in some detail by Friis.[14]

The dependence of the rate of polymerization on the emulsifier concentration is shown in Fig. 2 for both sodium lauryl sulfate and dioctyl sodium sulfosuccinate. The order of reaction was 1·0 and 0·9 respectively, much higher than that found with chemically initiated studies or with radiation initiation in flow systems. This point will be discussed later. The effect of dose rate on the polymerization rate is shown in Fig. 3 and found to be 0·7 order at 0°C. At higher temperatures the dependence increased reaching 0·9 at 50°C, close to that often observed with persulfate initiation. With a seeded

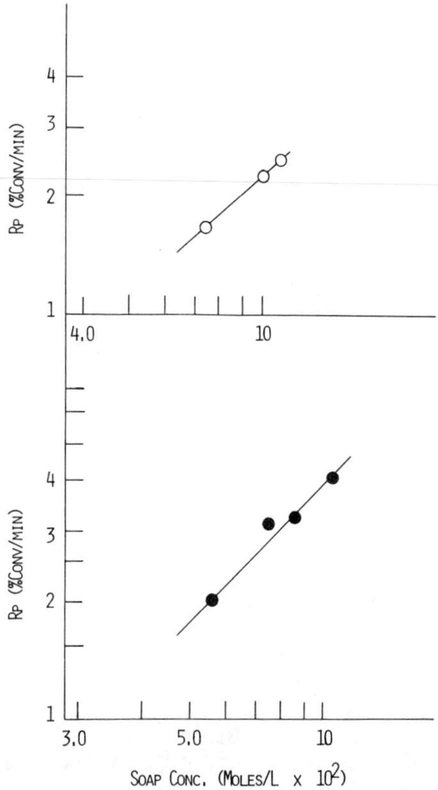

Fig. 2. Dependence of the polymerization rate for the radiation induced emulsion polymerization of vinyl acetate on the emulsifier concentration at 0°C. Dose rate 0·037 Mrad/h.[12] ○ Dioctyl sodium sulfosuccinate, slope = 0·9; ● sodium lauryl sulfate, slope = 1·0.

system the order of reaction dropped to 0·26 at 50°C. The data are included in Fig. 3. The rate itself with seeded systems was found to be of 0·7 order with the number of particles. Sunardi[15] recently reported that the polymerization rate decreased with increasing emulsifier content and increased with the dose rate to the 0·6–1·0 power with increasing emulsifier content.

The various orders of reaction rates are summarized in Table I, together with those found with persulfate initiated systems and those predicted by the Smith-Ewart theory. In general, the orders found with radiation initiation tend to approach the Smith-Ewart values more than the chemical initiated systems. There are considerable variations in the literature with persulfate

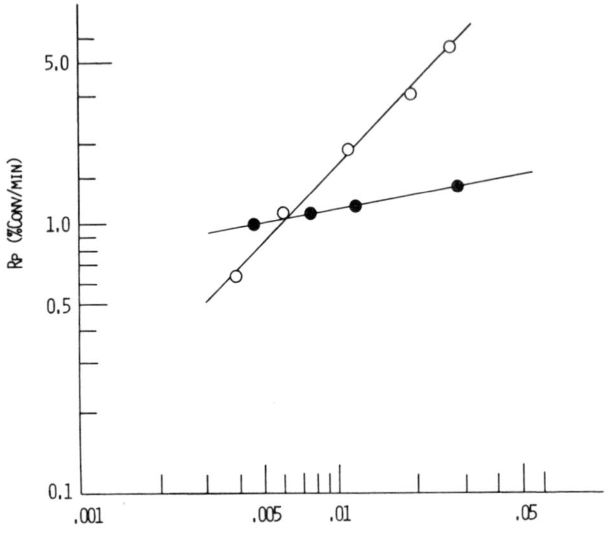

Fig. 3. Dependence of rate of polymerization on the dose rate for the radiation polymerization of vinyl acetate in emulsion at 50°C.[12] ○ Direct, slope = 0·9; ● seeded, fixed number of particles, slope = 0·26.

TABLE I
Orders of reaction for the rates with potassium persulfate and radiation polymerization

Variable	*Order* $K_2S_2O_8$	*Order* *Radiation*	*Order* *Smith–Ewart II*
Initiator or dose rate	0·5–1·0	0·5–0·9	0·4
Ditto with seeded system	0·8	0·26	0
Number of particles	0·2	0·7	1·0
Monomer–water ratio	0·35	0·2	0

Data taken from various sources.

initiation, however, as shown by the ranges reported in Table I. In one batch and two flow, micro pilot plant scale, studies[14,16] with radiation initiation, the orders with respect to emulsifier concentration were found to be approximately 0·3, close to the reported values found with persulfate initiation. The reasons for the discrepancy are not known. Similarly, the dependence on the dose rate has been found in other studies, both batch and flow were found to be close to 0·5 rather than the 0·7–0·9 order reported by Stannett *et al.*[12]

The kinetics of vinyl acetate polymerization are known to be very complex and a number of different theories have been advanced, these have been recently reviewed,[17] a new approach has also been presented by Litt and co-workers.[18] Furthermore, experimental results vary from worker to worker, probably due to changes in particle size and number because of coagulation and other reasons, and to the notably high sensitivity of vinyl acetate polymerizations to oxygen.

MOLECULAR WEIGHTS

Only a few definitive studies of the molecular weights appear to have been reported and in all cases only intrinsic viscosities were used for the measurements. Stannett *et al.*[12] found that essentially the molecular weights were independent of the dose rate (between 0·003 and 0·07 Mrad/h), the emulsifier concentration, with both sodium lauryl sulfate and dioctyl sodium sulfosuccinate, and the monomer–water ratio. O'Neill *et al.*[16] found the molecular weights to be 'relatively insensitive' to the dose rate and emulsifier content. Both authors found the molecular weight to increase with conversion but to decrease if the resulting polymers were hydrolyzed and reacetylated. This was a clear indication of branching due to chain transfer to the methyl groups. The results of O'Neill *et al.*[16] are depicted graphically in Fig. 4.

The studies by Sunardi[15] using intermittent radiation showed a linear decrease in the intrinsic viscosity with emulsifier content but only to a small degree, about twofold with a sixfold change in emulsifier content. Polyoxyethylene sorbitan monolaurate (Tween 20) was the emulsifier used and it is possible that some chain transfer took place.

Finally, the micro pilot plant flow system studies of Friis[14] showed the intrinsic viscosities to be independent of the emulsifier content and the number of particles. The values decreased slightly with increasing dose rate, 0·062 and 0·301 Mrad/h, probably due to some bimolecular termination at the higher dose rate. The effects were masked at higher conversions due to the extensive branching which is known to occur.

PILOT PLANT STUDIES

In addition to the basic laboratory studies described above, which were mainly with batch systems, two micro and one larger size pilot plants have

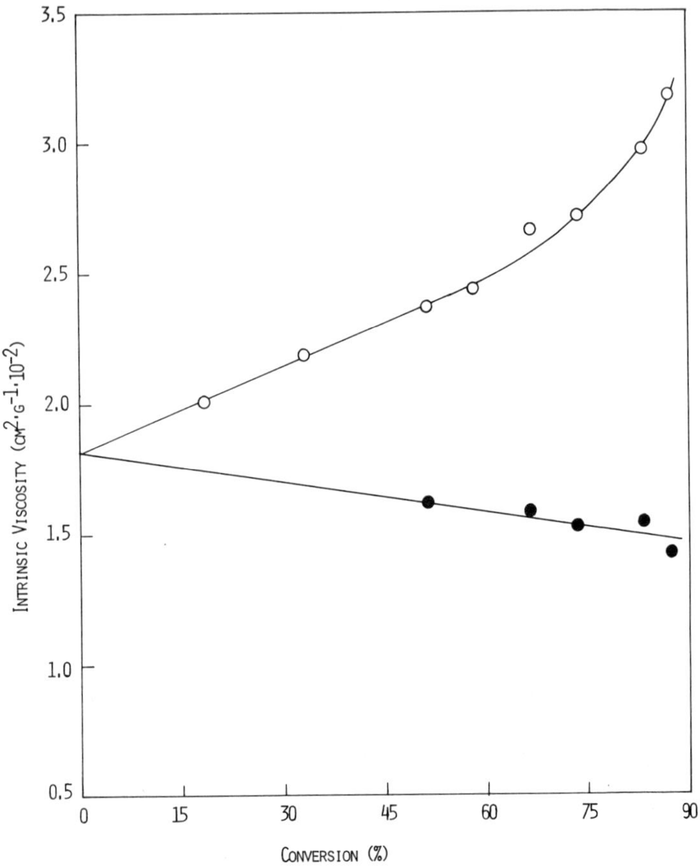

Fig. 4. Variation of the intrinsic viscosity of poly(vinyl acetate) produced by radiation polymerization in emulsion.[16] ○ Original; ● after hydrolysis and reacetylation.

been constructed and operated. All were based on the general principle of recirculation from a stirred vessel away from the radiation source. The emulsions were continuously circulated through tubing to an in-source 'plug-flow' reactor and back to the kettle. The two micro-pilot plants were largely based on the originally designed and operated pilot plants described by Stahel and co-workers.[19–21] The larger pilot plant constructed by Neutron Products Inc. was designed and constructed simultaneously but quite independently.[22] The radiation emulsion pilot plants described by Stahel and co-workers were investigated for styrene,[19] vinyl chloride–vinyl acetate copolymer[20] and for grafting to latex.[21] Recently the present authors have begun similar studies with vinyl acetate and the results will be published in due course.

The first micro pilot plant for vinyl acetate itself to be described was that of O'Neill *et al.*[16] and was essentially an all glass apparatus. The semi-continuous equipment shown in Fig. 5 consisted of a stirred, constant temperature 3000 cc glass reservoir from which the emulsion was pumped through a double helix 240 cc reactor situated within the radiation cavity. On leaving the reactor the emulsion flowed through a cooler and back into the reservoir. The reservoir has inlets for monomer, soap solution and pure nitrogen with which the apparatus was purged prior to and during the run. The 3000 cc charge of emulsion contained 2000 cc of sodium lauryl sulfate (SLS) aqueous solution and 1000 cc vinyl acetate. The operating conditions were: temperature (T) = 30° and 40°C, dose rate (I) = 87 krad/h, $(SLS)_{aq}$ between 0·05 and 5·0% wt/vol, flow rate (Q) = 0·9 liters/min, reactor residence time (t_r) = 0·26 min, and the stirring rate = 250 rpm; the initial charge to reactor volume ratio was 12·5.

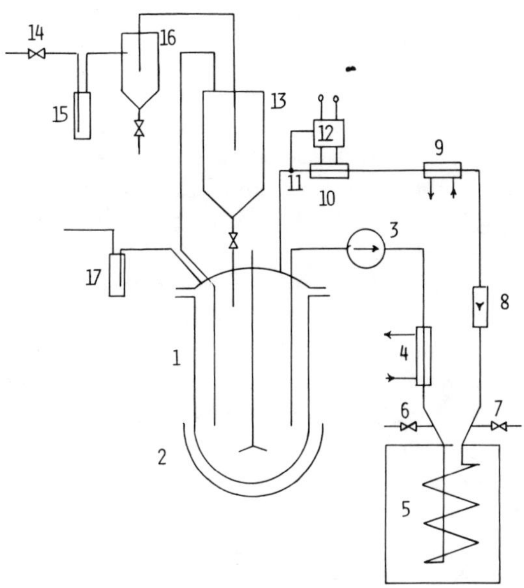

Fig. 5. Semi-continuous flow system for the radiation induced emulsion polymerization of vinyl acetate.[16] 1, Stirred reservoir; 2, heating mantle; 3, pump; 4, heat exchanger; 5, double helix reactor; 6, 7, sampling points; 8, rotameter; 9, cooler; 10–12, thermostat; 13, storage tank; 14, N_2 inlet; 15, humidifier; 16, separator; 17, N_2 outlet.

The conversion curves were found to be linear between 15 and 60%. No post effects were observed although this could have been caused by traces of oxygen in the system, due to diffusion of air through the thin walled polyethylene transfer lines. Increasing the flow rate decreased the rate of polymerization and at very fast rates of flow the polymerization ceased entirely. This effect again was ascribed to traces of oxygen in the equipment since the residence time in the oxygen free reactor itself became too short for the oxygen to be consumed. Some of the other results obtained by O'Neill *et al.*[16] have been described earlier.

The experimental equipment built by Friis[14] is shown in Fig. 6. There were two important differences from the O'Neill system. First, after initial purging of the system with nitrogen the polymerizations were conducted in the absence of any gas phase, and second the conversion–time curves were measured by means of a dilatometer rather than sampling. It was interesting that, in the absence of any oxygen, increasing the flow rate increased the rate of polymerization, confirming, perhaps, the explanation of O'Neill by the opposite effect of the flow rate. A very small post effect was found as discussed earlier together with other results of the Friis work. A full description of this work has been published in report form.[14]

The large scale pilot plant constructed by Neutron Products, Inc., was built as a model for a commercial scale facility.[22] Thus, it has features that would be important in an industrial process such as heat exchangers and pumps capable of handling high viscosity polymeric emulsions, multiple stage processing and the capability for continuous operation. Furthermore, all materials used in the latex formulations were of commercial grade with no purification prior to reaction.

The pilot plant consisted of four complete reactor loops, a typical one of which is shown in Fig. 7. The reactors varied in total volume from 3 to 35 gallons. They were all interconnected to permit multistage processing.

The reactants were pumped from the surge tank through a heat exchanger embedded in a six foot concrete floor which separated the radiation zone from the rest of the pilot plant. The reactants were then circulated around the source in a specified configuration, through another heat exchanger then back to the surge tank. All or part of the emulsion returning from the radiation zone could be drawn off for continuous operation.

The majority of the efforts of Allen *et al.*[22] at the Neutron products facility were directed toward producing a poly(vinyl acetate) latex suitable for production of high quality paints. They employed a viable commercial formulation, containing a comonomer, a complex emulsifier system, a stabilizer and a buffer.

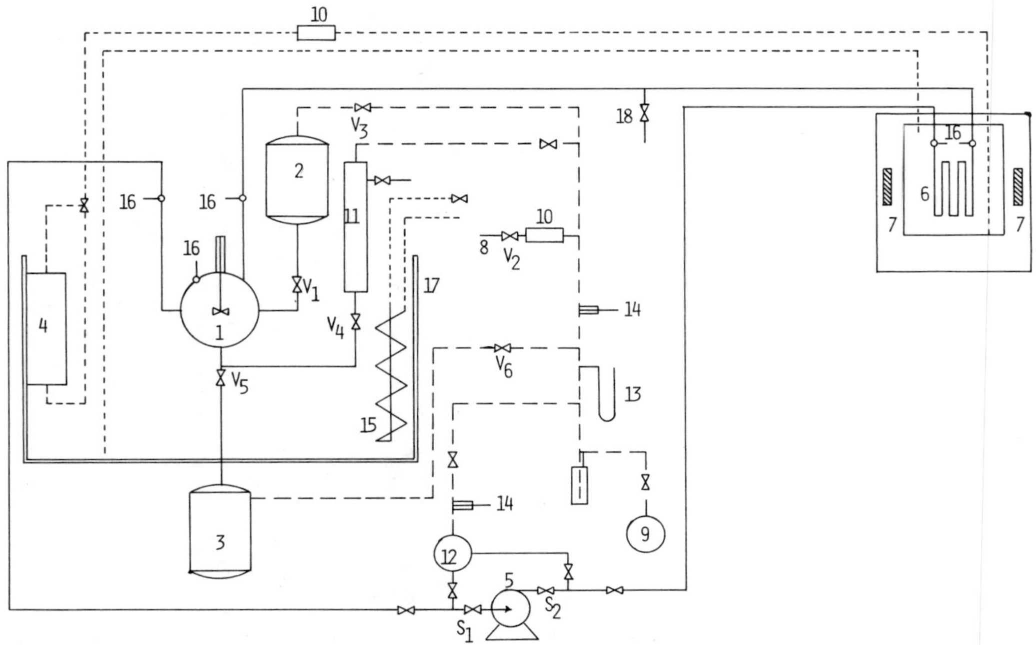

Fig. 6. Semi-continuous flow system for the radiation polymerization of vinyl acetate in emulsion.[14] —— Emulsion flow system; ---- temperature control circuit; – – – – N_2 and vacuum. 1, Backmix reactor; 2, feed tank, 3, receiver bottle; 4, thermostat; 5, circulation pump; 6, tubular flow reactor; 7, Co^{60}-source; 8, nitrogen supply; 9, vacuum pump; 10, flow meter; 11, dilatometer; 12, emulsion reservoir; 13, mercury manometer; 14, vacuum gauge tube; 15, cooling coil; 16, thermocouple; 17, water storage tank; 18, sample valve.

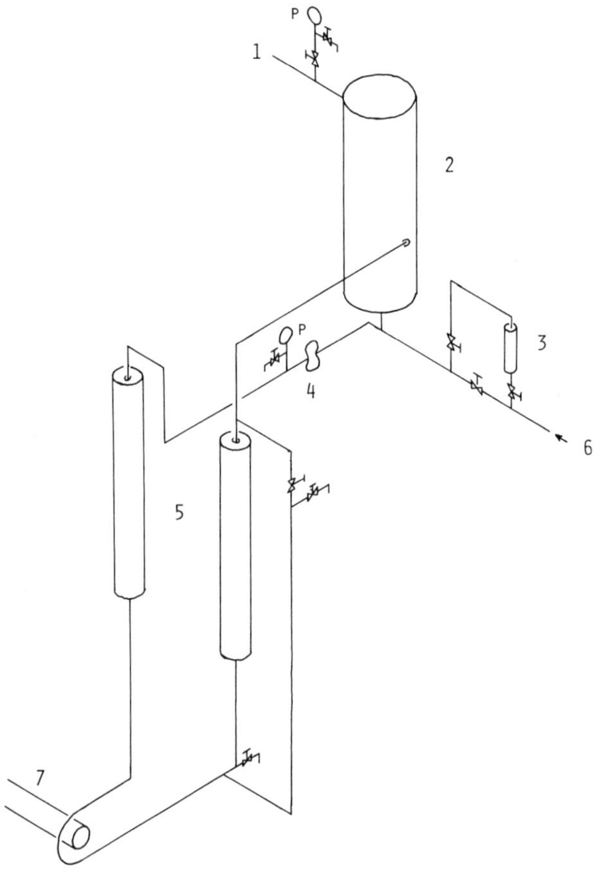

Fig. 7. Flow diagram for the pilot plant for the radiation induced emulsion polymerization of vinyl acetate (Neutron Products, Inc.[22]). 1, Vacuum; 2, surge tank; 3, rotometer; 4, loop pump; 5, heat exchangers; 6, feed; 7, radiation source; P, pressure gauge.

The highest quality latexes were produced with a two-stage batch or semi-batch process. The first stage, for initiation and propagation, was operated on either a batch basis or semi-batch basis with continuous monomer addition. Generally, the first stage was operated to about 12% residual monomer content at which time the product was transferred to the second stage for finishing under different conditions. The second stage was operated in either a batch or continuous manner. Experiments were also performed with the plant operating on an entirely continuous basis and a good quality latex was obtained.

The properties of the latexes and the paints produced from the latexes were evaluated by several commercial firms. The latex properties varied somewhat from sample to sample but, in general, the paints produced from the latexes were comparable to those produced from high-grade commercial latexes made by conventional means. A typical comparison is given in Table II.

TABLE II
Commercial comparison of latex from radiation pilot plant and a comparable commercial product†

	Radiation pilot plant	*Commercial vinyl acrylic*
Viscosity (Brookfield)	2450 cps	1000–2000 cps
% Solids	50·15	55·0
pH	5·9	4·6
Residual monomer	0·77	0·03
Loebel particle size	0·54	0·6–0·7
Intrinsic viscosity	1·35	0·7
Grit (ppm)	30	50–100
Freeze–thaw	failed first cycle	borderline
Interior latex paint properties		
Scrub-cycles	375	400
Enamel holdout	96·0%	80%
Color acceptance	excellent	excellent
Brushing	good	excellent
Lapping	excellent	excellent
Touch up	excellent	excellent

† Data from Neutron Products, Inc.[22]

One property of the paint produced from the radiation-catalyzed latex, enamel holdout, which is one measure of the ability of a paint to serve as a primer for enamel on unpainted wood, was superior to that of paint made from conventionally produced latex. This may have been due to the higher molecular weight obtained by radiation catalysis.

The results showed that production of latexes by radiation catalysis is a commercially feasible process. However, the authors did not appear to have resolved two process-development problems: the reduction of residual monomer to commercially acceptable levels, and the elimination of polymer build-up inside the process lines. The authors noted, however, that these are problems of the formulation and operating conditions rather than of the basic process itself.

Using the data obtained in the pilot plant, a conceptual design and cost analysis was made for a comercial-scale radiation catalysis plant. The conceptual design was based on a two-stage semi-continuous process. The first stage is operated on a semi-batch basis with continuous monomer addition and intermittent feed to the second stage which is operated on a batch basis. Economic analysis indicates that capital and operating costs for radiation catalysis and conventional catalysis plants would be about the same. But the authors felt that the greater enamel holdout of the radiation initiated product would justify the use of a radiation catalysis plant. In addition, the early work of Stannett *et al.*[12] showed that better wet-scrub resistance was obtained with paints formulated with high molecular weight radiation produced homo-vinyl acetate latexes compared with their commercial conventional counterparts.

The results of the laboratory and pilot plant studies show clearly that a radiation induced emulsion polymerization process for vinyl acetate is of considerable interest. In addition to the general advantages discussed in the introduction, the smooth and easy polymerization at low temperatures is of special interest with vinyl acetate and its copolymers since it leads to very high molecular weight products. High molecular weight polymerized acetate latexes are known to lead to paints and coatings with better ink and enamel holdout and superior wet-scrub resistance.

REFERENCES

1. P. E. M. Allen, J. M. Downer, G. W. Hastings, H. W. Melville, P. Molyneux and J. R. Urwin, *Nature,* **177**, 910 (1956).
2. S. Okamura, T. Motoyama, T. Manabe and H. Inagaki, *Large radiation sources in industry, Warsaw, 1959,* IAEA Vienna, Vol. I, 1960, pp. 368–73.
3. P. E. M. Allen, G. M. Burnett, J. M. Downer and H. W. Melville, *Makromol. Chem.,* **38**, 72 (1960).
4. H. Inagaki, K. Yagi, S. Saeki and S. Okamura, *Chemistry of High Polymers (Japan),* **17**, 37 (1960).
5. Ibid. **17**, 135 (1960).
6. K. Ohdan, H. Kamiyama and S. Okamura, *Ann. Report, Osaka Lab. for Rad. Chem. JAERI,* **5026**, 15 (1970).
7. M. Kitajima, H. Kamiyama and S. Okamura, Ibid. **5026**, 26 (1970).
8. H. Kamiyama, Ibid. **5029**, 33 (1974).
9. H. Kamiyama and K. Saito, Ibid. **5030**, 147 (1975).
10. H. Kamiyama and Y. Shimizu, Ibid. **5030**, 148 (1975).

11. H. Kamiyama, Ibid. **M. 6260,** 23 (1975).
12. V. T. Stannett, J. A. Gervasi, J. J. Kearney and K. Araki, *J. Appl. Polym. Sci.,* **13**, 1175 (1969). See also USAEC Report TID 2481, December 1, 1967.
13. D. Hummel, *Angew. Chem.,* **75**, 330 (1963).
14. N. Friis, Riso Report 282, Danish AEC, June, 1973.
15. F. Sunardi, *J. Appl. Polym. Sci.,* **24**, 1031 (1979).
16. T. O'Neill, J. Pinkava and J. Hoigné, *Proc. Third Tihany Symposium on Rad. Chem., Tihany, Hungary, 1971,* Vol. 1, 1972, p. 713.
17. V. T. Stannett, A. Klein and M. H. Litt, *Br. Polym. J.,* **7**, 139 (1975).
18. K. H. S. Chang, M. H. Litt and M. Nomura, this volume, Chapt. 6.
19. E. P. Stahel and V. T. Stannett, *Large radiation sources in industry, Munich, 1969,* IAEA Vienna, 1969, p. 135.
20. J. T. Tsai, E. P. Stahel and V. T. Stannett, *J. Appl. Polym. Sci.,* **23**, 2701 (1979).
21. K. Mitri, E. P. Stahel, T. Memetea and V. T. Stannett, *J. Macromol. Sci. Chem.,* **A.11**, 337 (1977).
22. R. S. Allen, J. A. Ransohoff and D. G. Woodard, USAEC Report ORO 673, October, 1969.

10

The Stabilization of Poly (Vinyl Acetate) Latexes by Polymethacrylic Acid

W. E. DANIELS,[a] J. W. VANDERHOFF,[b] C. T. ENOS,[a]
J. G. IACOVIELLO,[a] S. M. AHMED[c] and J. FROST[a]

[a] *Air Products and Chemicals Company, Allentown, Pennsylvania, USA;*
[b] *Emulsion Polymers Institute, Lehigh University, Pennsylvania, USA;*
[c] *Hercules Inc., Wilmington, Delaware, USA*

SUMMARY

It is known that polymethacrylic acid functions as a protective colloid for the emulsion polymerization of vinyl acetate. The resultant latex shows excellent mechanical freeze-thaw stability. The locus of the polymethacrylic acid has been determined by ion-exchange and conductometric titration, serum replacement using Millipore and Nucleopore filters, and exhaustive extraction of the dried films with hot water. The results show that 75–85% of the polymethacrylic acid is strongly adsorbed on the poly(vinyl acetate) latex particles and the remainder is solute polymer in the aqueous phase. The excellent stability of this latex is attributed speculatively to both steric and electrostatic repulsion arising from the adsorbed polymethacrylic acid.

INTRODUCTION

The emulsion polymerization (EP) of vinyl acetate (VAc) is unique in that it is readily stabilized by either steric, electrostatic, or combinations of these mechanisms.[1]

The mechanisms obtaining in electrostatically stabilized EP of VAc have been extensively studied by many workers. These efforts have been critically reviewed by Stannett *et al.*[2]

In contrast, sterically stabilized EP processes for VAc have received relatively scant attention in the scientific literature.

The majority of existing studies have dealt qualitatively with compositional effects of various stabilizers such as polyvinyl alcohol, hydroxyethyl cellulose, and nonionic surfactants[3–6] on latex properties such as viscosity and particle size distribution. A select few articles have dealt with the mechanistic details of sterically stabilized VAc EP reactions. Hartley[7] and Traane[8] have examined the fate of polyvinyl alcohol stabilizer in these processes, while Priest,[9] has reported on the effect of polymethacrylic acid on latex particle growth mechanisms.

Priest's study is of particular interest in that the action of polymethacrylic acid to generate large particle size was described, and its action compared with other protective colloids. The stability of these large (0·5–1·0) latexes was attributed to the presence of 'a hull of water soluble material which minimizes contact of the interior of the droplets on collision'. This statement is apparently one of the earliest mechanistic considerations on the role of protective colloids on stabilization of PVAc latexes, and perhaps one of the earliest enunciations of a currently widely held concept of steric stabilization. Further, it was allowed that 'it is not known whether these materials (i.e. protective colloids) are selectively adsorbed on the polymer particles or whether they may be actually chemically combined'.

This paper deals with the question of the fate of polymethacrylic acid in VAc polymerizations as revealed by determination of its locus in the final latex, and the nature of its bonding to PVAc latex particles.

EXPERIMENTAL

Polymethacrylic acid (PMAA)

PMAA was prepared essentially by the method of Priest. Methacrylic acid (polymerization grade, Rohm & Haas) was polymerized as a 10% w/w solution in water at pH 4·0, using NH_4OH as adjusting base at 50°C with the $Na_2S_2O_8/NaCH_2SO_2H$ FE^{+2} redox system. The final solution was clear, fluid, and contained 10·0% polymer solids. The pH was 4·5, and the polymer's viscosity [η] was 0·08 dl liter/g, as determined in 2N NaCl.

VAc emulsion polymerization

PVAc latexes with high solids content were prepared by the following technique: a 2·0 liter resin flask equipped with agitator, reflux condenser, thermometer, and N_2 purge facilities was charged with 961 g of the PMAA solution (pH 4·5). To this was added 14·0 g of the nonylphenolethoxylate ester of phosphoric acid (Wayfos D-10-N), 1·5 g of $Na_2S_2O_8$ dissolved in 50·0 ml water. The temperature was increased to 60°C while purging with N_2, and a delayed feed of 950 g VAc added continuously over a two hour period. The reaction was maintained by the simultaneous addition of a solution of 0·4 g sodium formaldehyde sulfoxylate in 75·0 ml of H_2O. The latex contained ~48·0% solids and its pH was 4·6.

The latex was filtered through a 325 mesh screen to remove coagulum. No residues were observed. The particle size distribution (PSD) is given in Fig. 1.

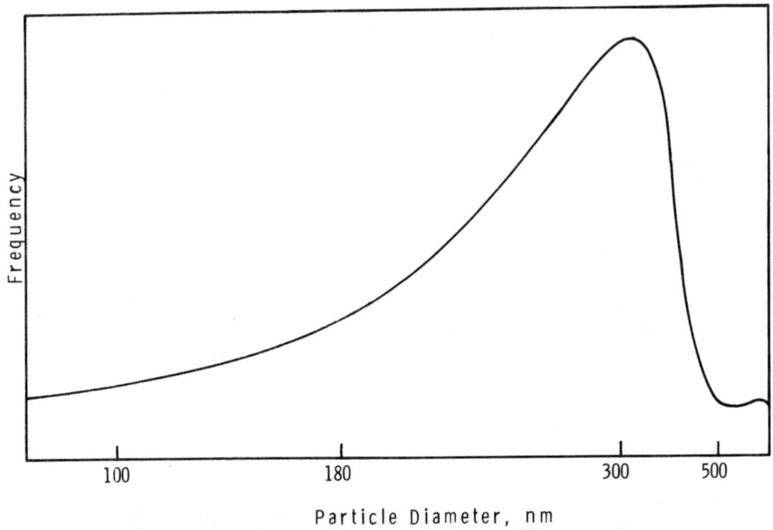

Fig. 1. Particle size distribution of latex 6 determined using the Joyce–Loebl disk centrifuge.

Particle size distribution

PSD was obtained by use of the Joyce Loebl disc photosedimentometer.

Ion exchange–conductometric titration of PMAA in latex

The procedure developed by Vanderhoff[10] was used. A modified method for determining rates of reaction of carboxylic acid species with ion exchange resin was also developed.

In this method, multiple samples of latex (5 g in 200 ml of water) were prepared and to each was added 20 g of purified cation form resin. The samples were gently agitated for two hours, then the resin removed by filtration through glass wool. Sample one was set aside, and to the remaining samples was added 20 g of mixed cation/anion exchange resins (5 g of cationic and 15 g of anionic resin). These samples were gently agitated for various times, filtered through glass wool, then titrated with a Mettler automatic titrimeter. *Molecular filtration* of the latexes was carried out using two techniques. The first is based on the use of Pellicon membrane (Millipore) designed for passage of 100 000 MW solute macromolecules. The second is based on serum replacement techniques[11] using Nucleopore filter media with uniform pore size.

Soxhlet extraction of latex films

Latex films about 5 ml thick were dried at room temperature, then at 50°C at reduced pressure. The dried film was powdered, extracted with deionized water for at least 48 hours, or until extractable matter became nil.

Latex mechanical stability

Latex stability was estimated by the standard Waring Blendor Test.[12]

Freeze–thaw stability

Standard procedures described elsewhere[13] were employed for these tests. Signs of failure noted were thickening, solidification, formation of sediment, or phase separation.

DISCUSSION

Polymerization

The VAc emulsion polymerizations using PMAA as protective colloid reported by Priest were not reproducible in stirred reactors at (monomer/water) phase ratios of 0·50. To provide for latex reactions with acceptable stability and solids content, it was found useful to employ about 2·0% by weight of the VAc of a phosphate ester anionic surfactant, Wayfos D-10-N. Results of several polymerizations as described in the experimental section are given in Table I.

Inspection of Table I reveals that neither the surfactant nor the PMAA was capable of maintaining colloidal stability during polymerization. The effect of PMAA at constant surfactant concentration appears to be directly proportional to the concentration of PMAA. Particularly notable is the effect of high levels of PMAA on reaction stability, as shown by the absence of coagulum, and by freedom from fouling of the reactor walls and agitator.

Physical properties of latex products

The particle size distributions of latex products 5 and 6 are given in Fig. 1. It is interesting to note that at the concentrations of PMAA examined, the particle size remains larger (0·3–0·5 μm) than expected for the level of

TABLE I
Stabilization of PVAc latex by polymethacrylic acid

Experiment No.	*% Surfactant*	*% PMAA*[a]	*Final % total solids*	*Final % coagulum*[b]	*Remarks*
1	0	0	—	—	Total coagulation
2	2·0	0	—	—	Total coagulation
3	2·0	1	~10·0	High[c]	
4	2·0	2·5	48·0	—	
5	2·0	5·0	48·0	None	
6	2·0	10·0	48·0	None	Reactor walls and agitator clean

[a] % Based on weight of VAc
[b] As retained on a 325 mesh standard sieve
[c] Latex nearly totally coagulated. Removal of coagulum by filtration through cheesecloth yielded very low solids content latex

anionic surfactant employed.[2,9] This is in keeping with the observations of Priest,[9] and with the known effects of protective colloids on VAc latex particle size.[3]

The mechanical stability of the products appears to be a direct function of PMAA concentration, as is freeze–thaw stability. These effects, given in Table II, are probably due to the presence of a larger envelope of water surrounding the particle, held in place by the layer of PMAA proposed by Priest and others.

TABLE II
Effects of PMAA on colloidal stability

Experiment No.	*% PMAA*	*Mechanical stability*	*No. of freeze–thaw cycles to failure*[a]
4	2·5	>30 min	—
5	5·0	>30 min	0
6	10·0	>30 min	>6

[a] Temperature +16°C

Latex characterizations

The high degree of colloidal stability exhibited by latexes prepared with high levels of PMAA occasioned renewed interest in the fate of PMAA used as protective colloid. It has been reported[1,7] that protective colloids become chemically grafted to PVAc particles through hydrogen atom abstraction in the high transfer activity of polymerizing VAc. It is also true that hydrogen atoms of the methacrylate α-methyl group participate sluggishly in transfer reactions. A detailed analysis of the locus of PMAA in these products would then serve to evaluate the relative importance of grafting or adsorption as the major stabilization mode in these reactions.

The locus of PMAA in latex sample No. 6 by the carboxyl distribution technique of Vanderhoff was evaluated under static conditions, and as a function of time of contact of the latex sample with the ion-exchange resin surface as described above. The results of the experiments are summarized in Table III.

TABLE III
Locus of PMAA by ion-exchange/conductometric titration

	meq COOH/g polymer	*Calculated*
Static		
1E[a] Dowex 50	1·01	1·1
1E Dowex 50	0·75	
Kinetic		
1E Dowex 50	1·0	1·1
1E No. 1 Mixed resin[b]	0·81	
1E No. 2 Mixed resin[b]	0·78	
1E No. 3 Mixed resin[b]	0·76	
1E No. 4 Mixed resin[b]	0·72	
1E No. 5 Mixed resin[b]	0·71	

[a] 1E = one exchange
[b] Two hours contact time/mixed resin exchange

It is clear that PMAA is removed from the latex by interaction with ion-exchange resin. A slow, continuous removal of PMAA, illustrated in Fig. 2, suggests that the fundamental source of stabilization is adsorption rather than grafting.

Further experiments were carried out to determine independently the amount of solute PMAA in the latex. Latex sample No. 6 was subjected to ultrafiltration employing a Millipore 'Pellicon' membrane. These membranes allow for selective passage of solute molecules as a function of their molecular size. Table IV summarizes these experiments.

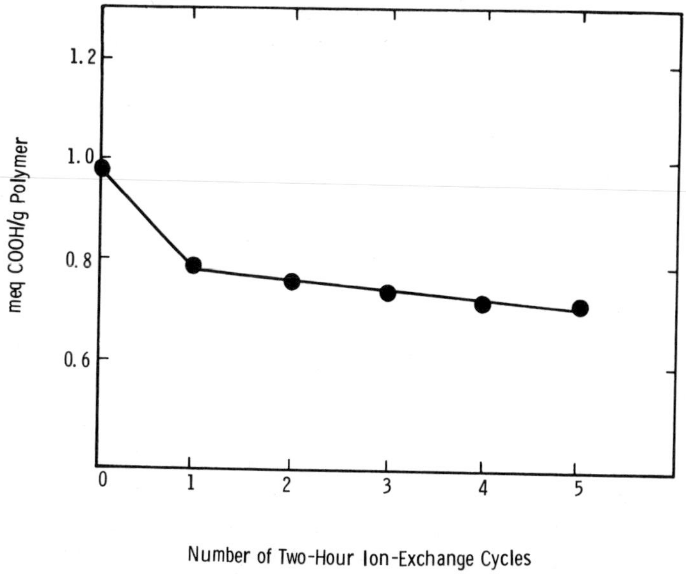

Fig. 2. Rate of removal of polymethacrylic acid from latex 6 particles by ion exchange.

TABLE IV
Molecular filtration of PMAA stabilized PVAc latex with 'Pellicon'[a] membrane

Latex sample[b]	*Eluant volume*[b]	*Cumulative g water solubles removed*[c]	*% of calculated water solubles removed*
200 ml	200 ml	3·86	36·1
	200 ml	4·77	44·5
	200 ml	5·18	48·7
	200 ml	5·48	51·2
	200 ml	5·71	53·3

[a] Millipore Corp.
[b] pH 9·1. Nearly identical results at pH 4·5
[c] Calculated total water solubles—10·7 g

The same sample was subjected to serum replacement filtration employing Nucleopore filters of uniform pore size. The results are summarized in Table V.

These filtration data also clearly show that PMAA is highly associated with the latex particles. The insensitivity of separation with high pH further indicates that the association of PMAA with particles is not due to hydrogen bonding of the type reported by Bailey *et al.*[14]

TABLE V

Ultrafiltration of PMAA stabilized PVAc latex[a] with Nucleopore[b] filters

	COOH/g polymers	*Calculated*
Ion exchanged/Dowex 50	1·01	1·1
Latex washed with water, ion exchanged/Dowex 50	0·53	
Collected latex washings	0·29	

[a] pH 4·5.
[b] Nucleopore Corp.

Details of the Millipore filtration are themselves of interest. First concentrated latex (48% solids) was found to very rapidly blind the membrane surface. Dilution to lower solids (200 ml water/100 ml latex) allowed for improved flow, which became impaired as latex solids approached the original value. Addition of fresh water improved flow again, but only until solids increased. This cycle was found to be indefinitely effective, and became the technique of filtration, resembling closely the serum replacement technique. The removal of PMAA by this technique is shown in Fig. 3 in which the

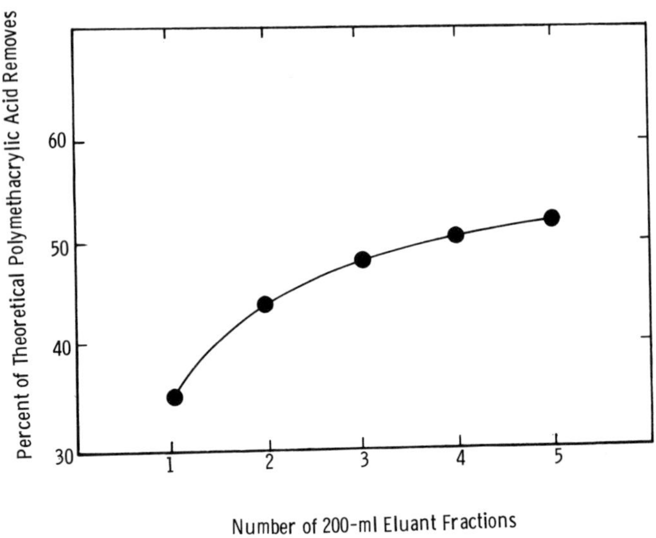

Fig. 3. Variation of rate of removal of polymethacrylic acid from latex 6 using serum replacement (Pellicon filtration membrane).

PMAA removal is shown as a function of the volume of water (relative to the starting latex volume of 200 ml).

Dry polymer films formed by normal drying of latex cast as a 2–3 mil wet film were observed to be very clear, but partially re-emulsifiable when exposed to water and rubbed with the finger. This behavior is not unusual in colloid protected latex films. It was unexpectedly observed that such films after heating at about 50°C under high vacuum became remarkably less sensitive to water, as shown by virtual elimination of re-emulsification on repetition of the above test.

Latex films dried under these conditions and then powdered by grinding were subjected to continuous extraction with water by the Soxhlet technique. It was observed that on prolonged extraction, PMAA was quantitatively extracted from these films. Furthermore, the given extracts were found to contain only PMAA, IR spectra indicating the absence of PVAc or surfactant. These data are presented in Table VI.

TABLE VI
Soxhlet extraction of dried latex solids[a]

Fraction	*Wt%*	*Identity*[c]
Water extractable	10·3 ± 0·6[b]	PMAA
Water insolubles	89·7	PVAc[d]

[a] 5·0 g powdered dried latex solids, extraction time four days
[b] 9·80 theory PMAA content
[c] IR, TGA
[d] ~40% soluble in THF

We interpret these data as being indicative of very strong adsorption of PMAA on PVAc particle surface, possibly involving hydrophobic bonding between the alpha methyl group of PMAA and the acetoxy substituents of the PVAc backbone, see Fig. 4. In the equilibrium state of the latex, it is likely that about 70–80% of the total PMAA is adsorbed on the particle surface with the balance present as free PMAA in the latex continuous phase. Carboxyl–acetate hydrogen bonded interactions are believed not to play a significant role in the bonding, as indicated by the high pH molecular filtration experiments. Grafting of PMAA by PVAc, if present at all, is evidently of a very low degree of importance.

The adsorbed PMAA layer contains both dissociated and undissociated carboxyl groups. Thus, in addition to the steric component of stabilization,

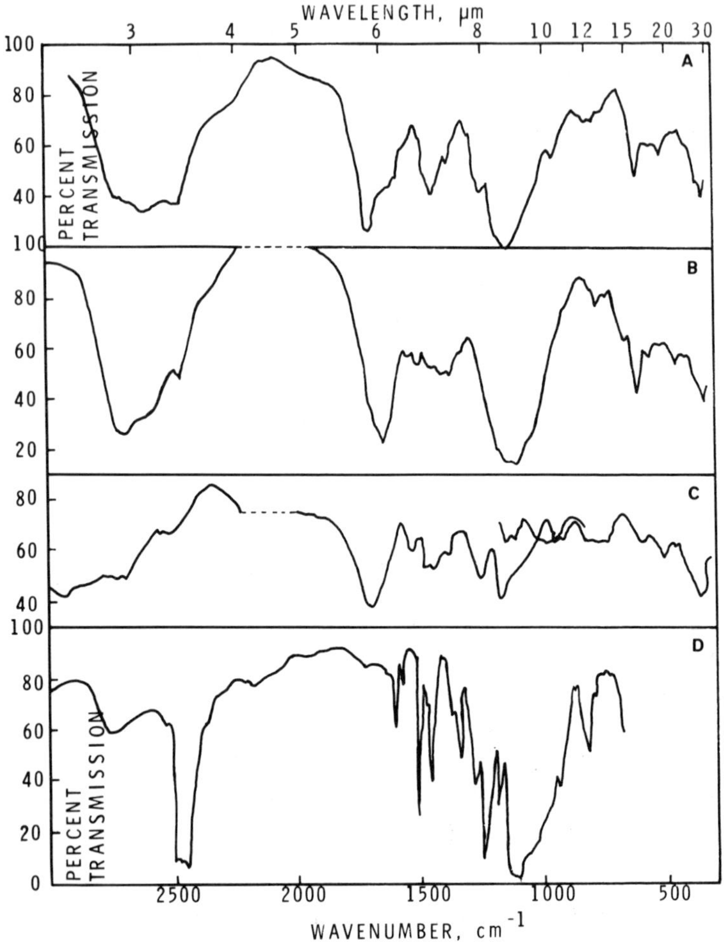

Fig. 4. Infrared spectra of dried film of latex 6 (A), extraction blank (B), polymethacrylic acid (C), and Wayfos DN-10 (D).

there will be a strong electrostatic repulsion between particles from carboxylate anion, as described by Napper and others. It is not known if PMAA materially influences the kinetics of VAc emulsion polymerization. It is tempting to speculate that aqueous solution polymerization of VAc should be sharply influenced by PMAA in that soluble PVAc oligo-molecules may diffuse to PMAA globular clusters[15] in which the processes of precipitation,

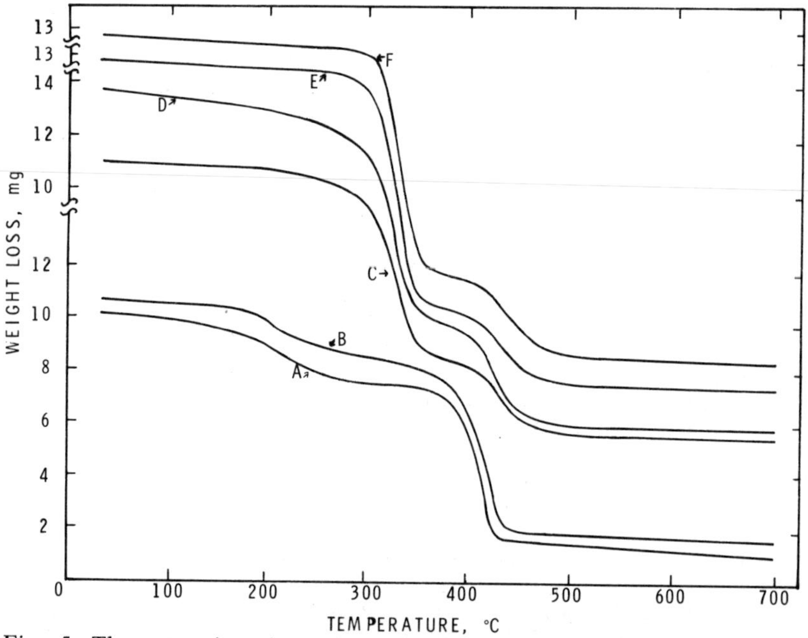

Fig. 5. Thermogravimetric analyses (under nitrogen; 10 psi; 250 cc/min) at 10°C/min: (A) polymethacrylic acid; (B) water-soluble extract of latex 6 film; (C) dried film of latex 6; (D) tetrahydrofuran extract of latex 6 film; (E) latex 6 film after extraction with water; (F) poly(vinyl acetate) standard.

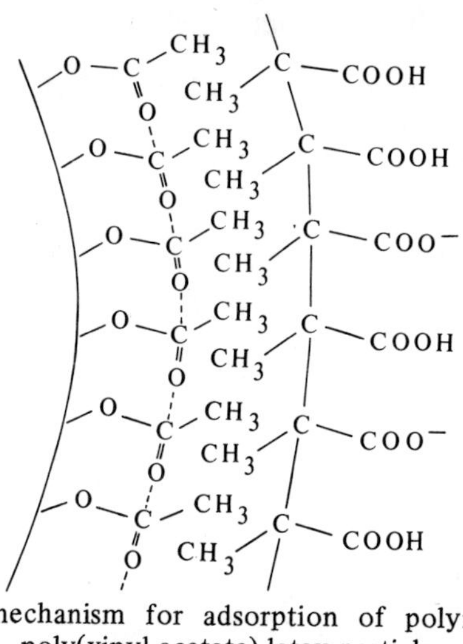

Fig. 6. Proposed mechanism for adsorption of polymethacrylic acid on poly(vinyl acetate) latex particles.

flocculation, and growth described by Priest could conceivably take place. This mechanism would predict a quasi-micellar role for PMAA, as opposed to an interceptory role in which flocculating PVAc particles are stabilized at some stage of growth by adsorption of PMAA (Fig. 5).

Future studies on these reactions will examine the mechanism of the reaction as revealed by kinetic and surface chemical studies on the adsorption of PMAA at various stages of particle growth (Fig. 6).

REFERENCES

1. M. K. Lindemann, *Encyl. Polym. Sci. Tech.*, **15**, 588 (1971).
2. V. T. Stannett, A. Klein and M. H. Litt, *Br. Polym. J.*, **7**, 139 (1975).
3. N. Platzer, *Mod. Plastics,* **23,** 95 (1951).
4. K. Noro, *Br. Polym. J.*, **2**, 128 (1970).
5. V. T. Shiriniyan, *Vysokomol. Soyed.*, **A-17**, 182 (1975).
6. D. French, *J. Polym. Sci.*, **32**, 395 (1958).
7. F. Hartley, *J. Polym. Sci.*, **34**, 397 (1959).
8. A. Traane, *J. Appl. Polym. Sci.*, **7**, 581 (1963).
9. W. J. Priest, *J. Phys. Chem.*, **56**, 1077 (1952).
10. J. W. Vanderhoff, *Polymer Colloids,* Plenum Press, New York, 1971.
11. S. M. Ahmed, M. S. El-Aasser, G. M. Pauli, G. W. Puehlin and J. W. Vanderhoff, *J. Coll. Interf. Sci.*, **73**, 388 (1980).
12. H. S. Stryker, A. F. Idelin and G. J. Mantell, *J. Appl. Polym. Sci.*, **10**, 81 (1966).
13. ASTM-D 2337 (27), 1973.
14. F. E. Bailey Jr., R. D. Lundberg and R. W. Callard, *J. Polym. Sci., H,* **2**, 845 (1964).
15. G. Barone, V. Crescenzi, B. Pispisa and F. Quadrifuglio, *J. Macromol. Chem.*, **1**, 761 (1966).

11

Particularities of Emulsion Polymerization of Vinyl Acetate in the Presence of Hydroxyethylcellulose

D. DONESCU,[a] KRISTIANA GOŞA,[a] I. DIACONESCU,[a] MAGDALENA MAZĂRE[a] and NICOLETA CARP[b]

[a] *Center for Plastic Materials Research,* [b] *Center for Physical Chemistry, Bucharest, România*

SUMMARY

Hydroxyethylcellulose (HEC) undergoes severe oxidative degradation reactions during the emulsion polymerization of vinyl acetate (VAc) initiated by potassium persulfate (KPS). In the presence of HEC the cleavage rate of KPS is greatly enhanced. The cellulose derivative forms a redox system with the initiator. The rate of cleavage is given by the equation:

$$-[KPS]_0/dt = K[KPS]_0\,[HEC]_0^{1/4}$$

The polymerization rate of VAc in the presence of HEC increases proportionally with monomer concentration. The reaction order to HEC depends on reaction conditions. The reaction order to initiator is 0·2. During emulsion polymerization of VAc, graft copolymers are obtained of the monomer onto HEC. This process was evidenced by viscometric measurements and UV spectra.

INTRODUCTION

From among protective colloids utilized for emulsion polymerization of vinyl monomers, hydroxyethylcellulose (HEC) is outstandingly efficient. Unlike most other colloids, during the polymerization reaction its macromolecule undergoes scission due to the initiator used.[1–3]

The investigation of the HEC–potassium persulfate (KPS), very often utilized in the technology of vinyl and acrylic emulsion, aims at finding the laws that rule the oxidative degradative process. The kinetic study of the emulsion polymerization of vinyl acetate (VAc) focuses on the effects of oxidative degradation and of other factors on the polymerization rate and on the structure of the polymers obtained.

EXPERIMENTAL

Hydroxyethylcellulose was a Natrosol 250 LR, Hercules Power Co. product. It is a cellulose derivative of 2·5 molar substitution degree.

Potassium persulfate p.a. reagent grade, with no additional purification was used.

Vinyl acetate was used after rectification.

Working parameters: Temperature, 70°C for degradation of HEC and 65°C for polymerization. No precautions for removal of oxygen were taken.

HEC oxidative degradation was studied by viscometric and UV spectrophotometric methods.

The viscosity was determined by an Ubellhode viscometer. The solutions employed contained $NaHCO_3$ (the weight ratio $NaHCO_3$/KPS was 1/1).

UV spectra were recorded by Specord UV-VIS, C. Zeiss Jena, in cells of 0·2 cm. Initial HEC solution (without KPS) was employed for compensation. The variation of absorbance (E) at 263 nm was studied. At this wavelength the carbonyl groups following oxidative degradation show maximum absorption.[4–6]

For the benzene-soluble fraction of polymer differential spectra were used (solutions at different temperature and solutions at 20°C were introduced in cells). The solvent was an ethanol/water (90/10 vol.) mixture. The concentration of the solution was 0·032% grav.

The time variation of KPS concentration was determined by iodometric titration.

Transmission electron micrographs were obtained on the osmium tetroxide-stained films of benzene-soluble fractions. The films were casts from benzene solutions of 0·001% grav. concentration made directly on the copper grids. The morphology of these films was studied after a staining time of 14 hours. Only HEC units are stained by this technique.

THE DEGRADATION OF HEC

After KPS addition the viscosity of HEC solution, or the ratio viscosity/initial viscosity exhibits a drop and after some time a constant value (Fig. 1).

The decrease of the viscosity of HEC solution following KPS addition, points to the fact that macromolecules undergo an oxidative degradation, which entails their scission.

Following oxidative degradation, carbonyl groups are formed on macromolecules of cellulose derivatives.[4–6] The carbonyl groups show maximum UV absorption at 263 nm.

A plot of the UV absorbance of HEC solutions at 263 nm versus concen-

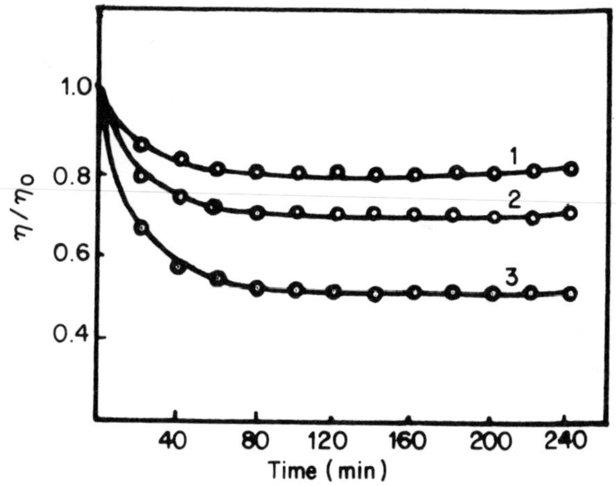

Fig. 1. The time variation of the ratio of instantaneous over initial viscosity (η/η_0) for various initial concentrations. $[HEC]_0 = 3$ g/liter (1), 5 g/liter (2), 10 g/liter (3) ($[K_2S_2O_8]_0 = 1$ g/liter).

tration shows that absorbance increases with HEC concentration. The variation of final absorbance with initial HEC concentration is not linear. The value of absorbance over initial HEC concentration ratio is a measure of oxidative degradation occurring per arbitrarily chosen HEC unit. The $[K_2S_2O_8]_0/[HEC]_0$ ratio represents the amount of initiator corresponding to the unit of HEC. Deviation from linearity of the curve in Fig. 2 shows clearly that in the case of large KPS amounts per HEC unit, the efficiency of oxidation decreases. In these instances the efficiency of KPS–HEC interaction diminishes as the number of radicals involved in secondary reaction increases.

The concentration of HEC exerts a very strong influence upon the decomposition reaction rate of KPS as well (Fig. 3).

As was specified for macromolecular alcohols (polyvinyl alcohol)[7] or small molecular alcohols[8,9] the rate of decomposition of KPS rises strongly in their presence. Unlike the polyalcohols, the initial decomposition rate of KPS depends on $\frac{1}{4}$th power of initial HEC concentration (calculated from data plotted in Fig. 3). In fact the shape of the curve in Fig. 3 points to the formation of a KPS–HEC redox system; the reducing part consists of hydroxyl groups in the cellulose derivative molecule.

The initiator concentration also affects the oxidative degradation of HEC (Fig. 4).

Figure 4 shows a plot of final absorbance versus initial initiator concentration. The dependence is not linear. The increase of initiator concentration causes this efficiency to lower, probably due to the increase of the relative importance of transfer reactions undergone by radicals.

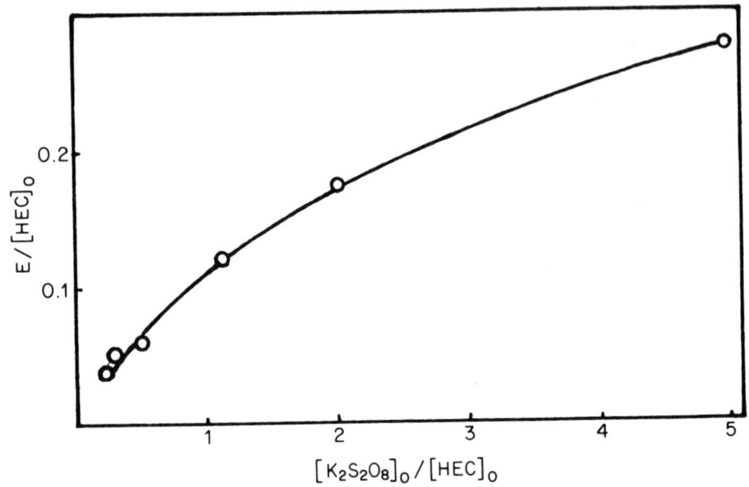

Fig. 2. The variation of absorbance (E)/$[HEC]_0$ versus initial HEC concentration ($[K_2S_2O_8]_0$ = 5 g/liter, conversion of $K_2S_2O_8$ = 80%).

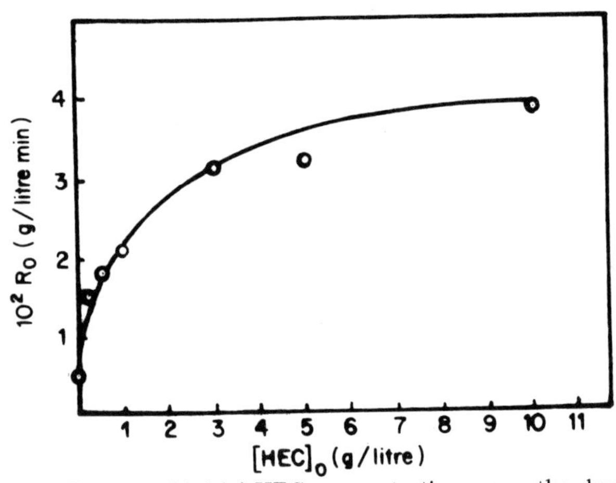

Fig. 3. The influence of initial HEC concentration upon the decomposition rate (R_0) of $K_2S_2O_8$ ($[K_2S_2O_8]_0$ = 5 g/liter).

For 2·5 g/liter KPS consumed, the UV absorbance decreases with initial KPS concentration, which supports the above statement (Fig. 4 curve 2). This phenomenon agrees very well with the data in Fig. 2 but not with the polyvinyl alcohol–KPS system. In this latter case the concentration of carbonyl groups is directly proportional to KPS concentration.[7]

The increase of initiator concentration entails a proportional increase of its decomposition rate (Fig. 5).

The dependence of KPS decomposition rate on its concentration is similar to that found in investigations on KPS reaction with other alcohols.[7,10]

The values of initial KPS decomposition rate at 60–85°C have allowed us

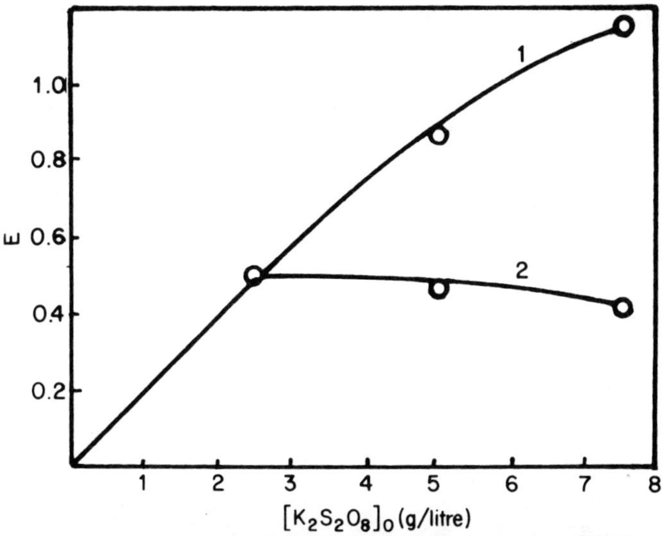

Fig. 4. Absorbance (E) versus initial $K_2S_2O_8$ concentration: (1) E_{final}; (2) E for 2·5 g/liter consumed $K_2S_2O_8$.

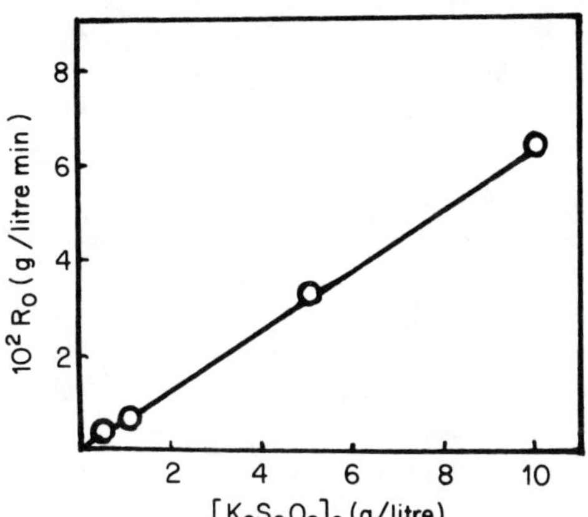

Fig. 5. The influence of $K_2S_2O_8$ concentration upon its initial decomposition rate ($[HEC]_0$ = 10 g/liter).

to calculate the activation energy. It amounts to 60 kJ/mole, a value close to that determined in reactions with isopropanol.[11] This value is lower than that determined in the KPS decomposition reaction in water.[11] This result lends further support to the idea that an HEC–KPS redox system is formed.

THE EMULSION POLYMERIZATION

Due to this degradative process of the protective colloid the kinetics of emulsion polymerization of VAc in the presence of HEC has certain distinctive aspects.

The dependence of maximum polymerization rate on initial monomer concentration is plotted in Fig. 6.

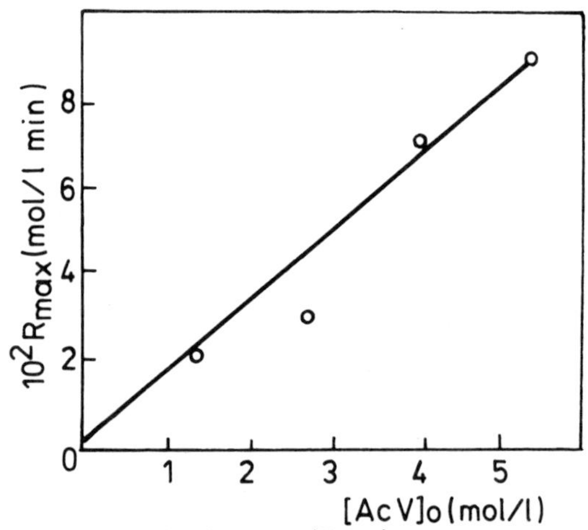

Fig. 6. Maximum polymerization rate (R_{max}) versus monomer concentration ($[K_2S_2O_8]_0 = 1{\cdot}15 \times 10^{-2}$ mole/liter, $[HEC]_0 = 4{\cdot}37 \times 10^{-2}$ mole structural units/liter).

As in cases where polyvinyl alcohol or ethylene oxide–propylene oxide block-copolymers are employed, the plot is a straight line not passing through the origin.[12–14]

The sudden increase of polymerization rate above the solubility limit of VAc when emulsified monomer particles are present may have two main causes: the increase of KPS decomposition rate in presence of existing monomer at saturation concentration in aqueous phase, or the active macroradicals have a rich reservoir of monomer particles or monomer shell around polymer particles. The first alternative is ruled out since the decomposition rate of KPS in presence of HEC is not affected by initial monomer concentration.

The maximum polymerization rate of VAc also increases with KPS concentration (Fig. 7).

The reaction order for KPS is therefore 0·2. This is the lowest reaction order we have found for emulsion polymerization of VAc, lower than that for other emulsifiers.[12–14] The results in Fig. 7 may be accounted for as follows: KPS forms a redox system with HEC which supplies a large number of radicals from the beginning of polymerization. Since the initiation rate is limited by reaction conditions, the additional increase in the number of radicals formed does not significantly affect the polymerization rate.

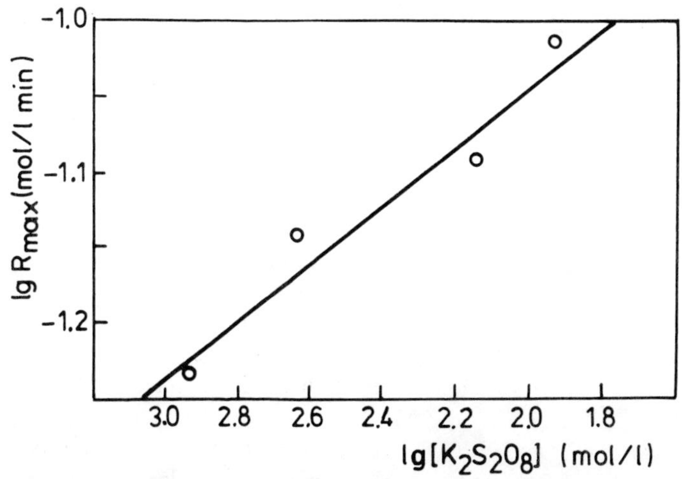

Fig. 7. The maximum polymerization rate–initial initiator concentration correlation ($[VAc]_0 = 4{\cdot}07$ mole/liter, $[HEC]_0 = 4{\cdot}37 \times 10^{-2}$ mole/liter).

The dependence of polymerization rate on colloid initial concentration is shown in Fig. 8.

Examination of Fig. 8 shows that the reaction rate increases with initial HEC concentration. This increase may be due to the increase of decomposition rate of KPS and of the number of particles with the increase of initial HEC concentration. In the polymerization conditions of Fig. 8 the initiator decomposition rate rises as HEC concentration increases.

At a 7·6–1·9 ratio of HEC concentration to KPS concentration as in Fig. 8, the efficiency of KPS in oxidative degradation of HEC is lower. A tenfold decrease of KPS concentration induces a decrease of polymerization rate with the increase of HEC concentration.

In these conditions the efficiency of the HEC–KPS interaction rises, which causes the number of carbonyl groups formed to increase. On transfer to

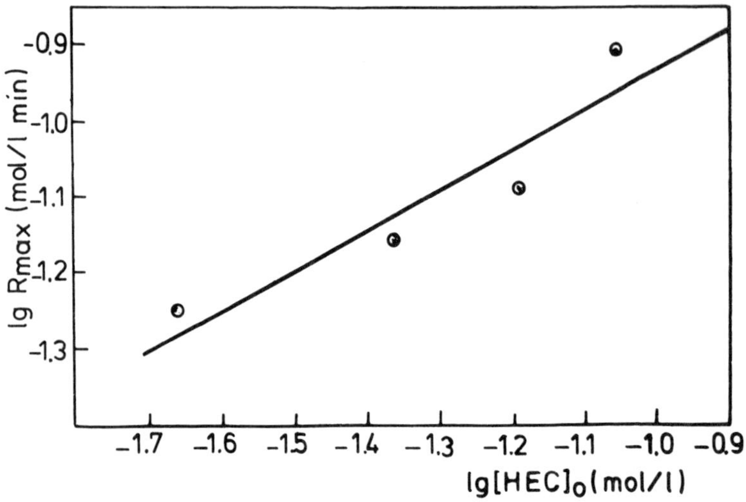

Fig. 8. The influence of initial HEC concentration upon maximum polymerization rate ($[VAc]_0$ = 4·07 mole/liter, $[K_2S_2O_8]_0 = 1{\cdot}15 \times 10^{-2}$ mole/liter).

surfactant, either to carbonyl groups or to unaltered molecules, many fewer reactive centers are formed. Therefore it turns out that, like polymerization in the presence of polyvinyl alcohol and ethylene oxide-propylene oxide block-copolymers, the reaction order for surfactant depends on the reaction conditions.

The existence of an HEC–KPS redox system proves that the formation of free radicals on HEC chains is possible. These free radicals may act as growing centers, which on VAc addition yield graft-copolymers. As oxidative degradation is accompanied by the scission of the chain, free radicals are formed at the ends of HEC fragments as well. VAc addition to these radicals may cause formation of block-copolymers.

All the considerations above suggest that during emulsion polymerization of VAc in the presence of HEC, graft-copolymers may be formed whose structure may be that of block-copolymers.[15]

The existence of graft-copolymers is also supported by the viscometric behavior of benzene-soluble polymer fractions (Fig. 9). The variation of intrinsic viscosity with temperature is non-monotonous, as was also shown for VAc polymerization in the presence of polyvinyl alcohol and ethylene oxide–propylene oxide block-copolymers.

The UV spectra of benzene-soluble fractions also show a non-monotonous variation of maximal absorbance with temperature (Fig. 10).

This is evidence for formation of HEC(PVAc) graft-copolymers.

Transmission electron micrographs show that the films of benzene-soluble fractions cast from benzene at room temperature have heterogeneous structure (Fig. 11(a)). The darker regions are HEC separate domains in the continuous PVAc phase.

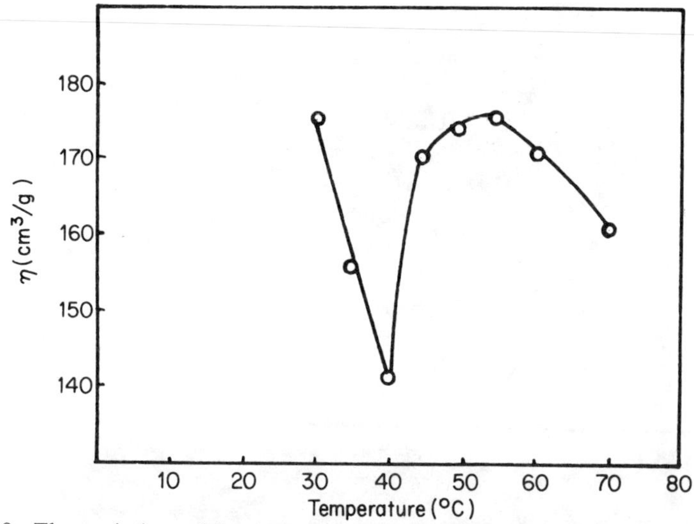

Fig. 9. The variation of intrinsic viscosity (η) of benzene-soluble fractions versus temperature ($[VAc]_0 = 4{\cdot}07 \times 10^{-2}$ mole/liter, $[HEC]_0 = 4{\cdot}37 \times 10^{-2}$ mole/liter, $[K_2S_2O_8]_0 = 1{\cdot}15 \times 10^{-2}$ mole/liter).

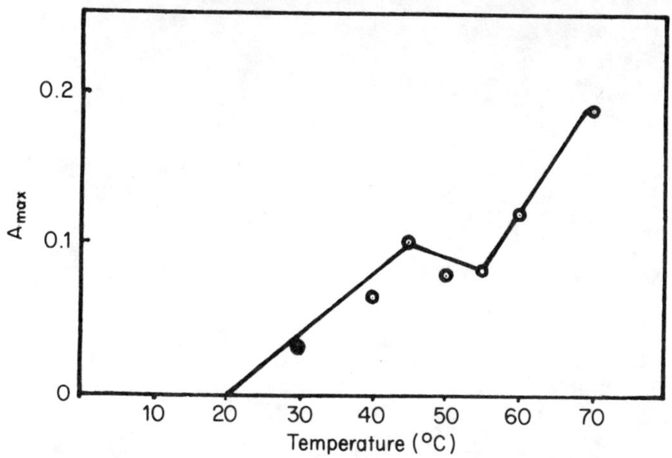

Fig. 10. The variation of maximum absorbance (A_{max}) in the UV spectrum of benzene-soluble fractions versus temperature (conditions as in Fig. 9).

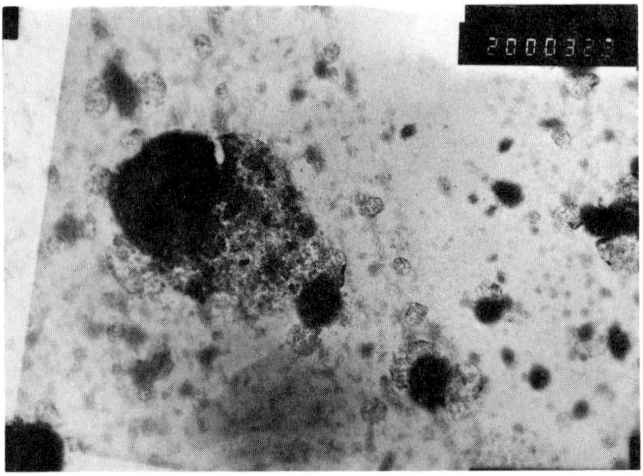

(a)

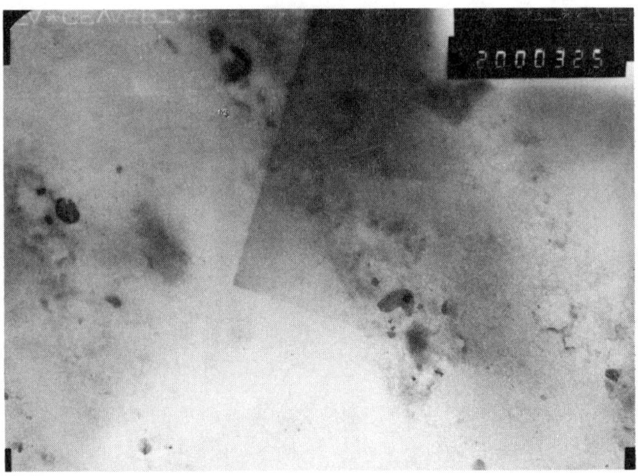

(b)

Fig. 11. Transmission electron micrographs of osmium tetroxide-stained films of benzene-soluble fractions. (a) At room temperature (20°C); (b) treated at 65°C for 60 min. (Conditions as in Fig. 9; VAc conversion = 20%; × 20 000.)

When the films are treated at 65°C the structure of the films becomes homogeneous (Fig. 11(b)). Above 65°C the incompatibility of hard HEC and soft PVAc segments vanishes.

In conclusion, emulsion polymerization of VAc in the presence of HEC is affected by the oxidative degradation of the protective colloid produced by initiator. During polymerization the monomer is grafted on the macromolecule of cellulose derivative.

ACKNOWLEDGEMENT

The authors wish to thank Mr J. Languri for the transmission electron micrographs.

REFERENCES

1. G. Harris, *Amer. Paint J.*, **13**, 14 (1972).
2. C. Bondy, *J. Oil Col. Chem. Assoc.*, **49**, 1045 (1966).
3. C. E. Reader, *Paint Technol.*, **32**, 14 (1967).
4. Z. A. Rogovin and N. M. Sarighina, *Chimia Celulozei şi a Substanţelor Insoţitoare,* Technică, Bucharest, 1958.
5. A. Bas, *J. Appl. Polym. Sci.*, **16**, 2567 (1972).
6. N. A. Ghanem, N. I. El-Avady, K. Singer and M. I. Poly, *Europ. Polym. J.*, **15**, 1007 (1978).
7. Y. Ikada, Y. Mishizaki and I. Sakurada, *J. Polym. Sci.–Polym. Chem. Ed.*, **12**, 1829 (1974).
8. J. O. Edwards, A. R. Galopo and J. E. McIsaac, *J. Amer. Chem. Soc.*, **88**, 3841 (1966).
9. J. M. Kolthoff, E. J. Meehan and E. M. Carp, *J. Amer. Chem. Soc.*, **75**, 1439 (1953).
10. D. L. Ball, M. Crutchfield and J. O. Edwards, *J. Org. Chem. Soc.*, **25**, 1599 (1960).
11. A. S. Dunn and C. J. Tonge, *Polymer Preprints,* 1261 (1977).
12. V. Dimonie, D. Donescu, M. Munteanu, C. Hagiopol and I. Gavăt, *Rev. Roumaine Chim.*, **19**, 503 (1974).
13. I. Gavăt, V. Dimonie, D. Donescu, C. Hagiopol, M. Munteanu, K. Goşa and Th. Deleanu, *J. Polym. Sci. Polym. Symp.*, **64**, 125 (1978).
14. D. Donescu, K. Goşa, M. Munteanu and I. Gavăt, *Rev. Roumaine Chim.*, **24**, 1399 (1979).
15. J. Cl. Mileo and L. Nicolas, *J. Appl. Polym. Sci.*, **11**, 425 (1967).

12

Preparation, Characterization and Properties of Vinyl Acetate – Butyl Acrylate Copolymer Latexes

M. S. EL-AASSER, T. MAKGAWINATA, S. MISRA and J. W. VANDERHOFF
Emulsion Polymers Institute, Lehigh University, Pennsylvania, USA
and
C. PICHOT and M. F. LLAURO
Laboratoire des Matériaux Organiques, CNRS, Villeurbanne, France

SUMMARY

In the field of emulsion polymerization, it is well known that the nature of the process (batch, semi-continuous, or continuous methods) can affect the kinetic behavior (rate of polymerization) or some polymer latex properties (number and size of particles, particle size distribution), and molecular weight development.[1–3] *In the particular case of emulsion copolymerization the homogeneity of the copolymer composition is also found to be controlled by the process.*[4] *In fact, a copolymer composition-drift generally prevails, due to change in the composition of the monomeric mixture versus conversion, as a result of differences in reactivity ratios and/or water solubility between the two monomers. Obviously, the compositional homogeneity of the copolymer products determines their molecular properties and consequently their physical and mechanical behavior.*

The homogeneity of the copolymer composition within the latex particle is expected to have a strong effect on the surface and colloidal properties; however, relatively less amount of work was devoted to this aspect. Vinyl acetate copolymer latexes are actually widely prepared using semi-continuous methods to improve the homogeneity of the polymers. Chujo et al.[5] *have already investigated the effect of monomer addition methods on the homogeneity of 70/30 (wt%) vinyl acetate–butyl acrylate copolymer latexes and showed that heterogeneous copolymer compositions were obtained by total monomer addition at the beginning of the polymerization.*

The purpose of the current work is to examine systematically the effect of the comonomer composition and the method of monomer addition (two factors which do affect the compositional homogeneity) on the bulk and colloidal properties of vinyl acetate–butyl acrylate copolymer latexes. The present review aims at giving detailed results on preparation of latexes (kinetic investigations), molecular characterization of the copolymers (molecular weight distribution, dynamic mechanical behavior, homogeneity of copolymer composition and sequence distribution); and colloidal properties of the copolymer latexes (particle size and distribution, surface characterization, ageing effect, and stability studies).

EXPERIMENTAL

Materials

Both monomers, vinyl acetate and butyl acrylate, were distilled under vacuum to remove the inhibitor. Deionized and deoxygenated water was used in all polymerizations. Ammonium persulfate and sodium bicarbonate were reagent grade (Fisher); Aerosol OT and Aerosol MA (80%) (American Cyanamid Co.) were used as received.

Preparation of copolymer latexes

The latexes were prepared by emulsion polymerization using recipes given in Table I. The conventional batch technique (bottle polymerization) was used to prepare latexes of Series A. In this case all ingredients except the initiator were charged into the bottles and then purged with laboratory grade nitrogen. These bottles were tumbled end-over-end at 35 rpm, in a constant-temperature bath at 75°C, for one hour prior to the initiator injection and for 24 hours thereafter. A semi-continuous process of monomer addition was used to prepare latexes of Series B. The polymerization was carried out in a 500 ml four-necked flask immersed in a constant temperature bath at 75°C and equipped with a reflux condenser, a two-bladed stainless steel stirrer, and two 5 ml graduated dropping funnels. Deionized water, emulsifiers and buffer were initially charged and maintained under constant agitation of 100 rpm under nitrogen atmosphere for half an hour followed by an approximate addition of 10% of the total ammonium persulfate initiator (dissolved in deionized water). The monomer mixture and the remainder of the initiator were simultaneously added, through dropping funnels, at rates of 0·30 ml/min and 0·40 ml/min, respectively.

Thus, under these conditions, all the initiators were introduced in three hours and the monomer mixture in 4·5 hours. A high degree of conversion could be obtained by carrying all polymerizations for only 5·5 hours. The

TABLE I
Polymerization recipes

Monomer mixture = 100 g
Deionized water = 200 g
Aerosol OT† = 1·5 g
Aerosol MA† = 1·5 g
Sodium bicarbonate = 0·3 g
Ammonium persulfate = 0·3 g

		Final percent conversion			
Butyl acrylate content		*Batch process*		*Semi-continuous process*	
Weight %	*Mole %*	*Sample (series A)*		*Sample (series B)*	
0	0	A.1	100	B.1	100
15	11	A.2	100	B.2	100
23·5	18		—		91
37·0	29	A.3	100	B.3	92
50	40·5		—		96
60	51	A.4	99·7	B.4	97
100	100	A.5	100	B.5	100

† American Cyanamid Co.

latexes were filtered and residual monomer, if any, was removed by steam distillation under vacuum with a Fisher Flash Evaporator. The final conversion, reported in Table I, was measured gravimetrically and by gas chromatography. Kinetic studies were performed through gas chromatography analysis of the unreacted monomer mixture using the automatic device already described[6] and given in Fig. 1.

Molecular characterization

Molecular weight and distribution

The average molecular weight and the molecular weight distribution of each copolymer latex was determined by gel permeation chromatography (GPC) using a Waters Associates 201 A 6PC equipped with a differential refractometer detector. Four microstyragel columns were used with pore size 10^3, 10^4, 10^5 and 10^6 Å. All copolymer samples were dissolved in tetrahydrofuran (THF) at concentrations of 0·1 to 0·2 wt%. For bimodal distribution,

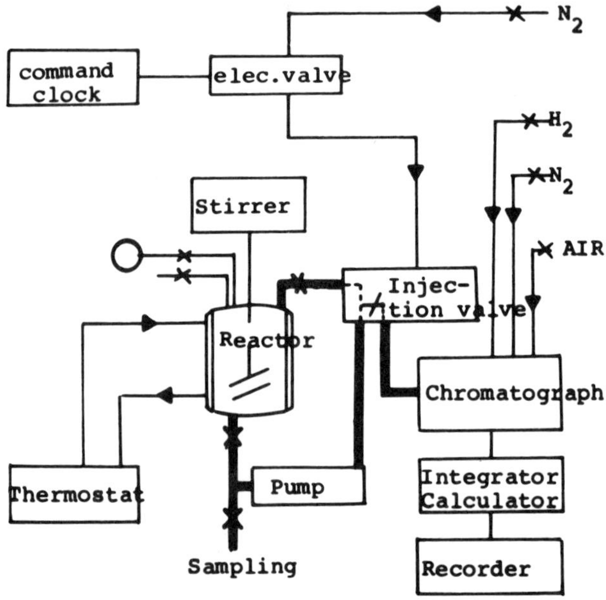

Fig. 1. General scheme of polymerization apparatus for kinetic studies.

a simulation computer program was used to obtain molecular weight values corresponding to each peak.

Dynamic mechanical properties

Films of latexes, 0·3 to 0·4 mm thick, described in Table I were dried at room temperature (*ca.* 22°C) in glass petri dishes pretreated with a mold release agent for epoxy resins (Parfilm; Price-Driscoll Corp.). These films were used for mechanical characterization after ageing for 40–50 days at room temperature, a sufficient time for completion of further gradual coalescence.[7, 8]

A Rheovibron viscoelastometer, Model DDV-II (Toyo Measuring Instrument Co.) was used to measure the storage modulus (E'), the loss modulus (E''), and the loss tangent ($\tan\delta$). All measurements were made at 110 Hz, over a temperature range from −90°C to 10°C above the glass transition temperature (*Tg*), and at a heating rate of 1°C/min. Icing of the specimens at low temperatures was avoided by sweeping dry nitrogen through the chamber. *Tg* for all samples was determined from the peak on the loss modulus spectrum.

A differential scanning calorimeter, DSC 1-b (Perkin-Elmer Corp.) was also used to measure the glass transition temperatures of the above samples. 10–20 mg specimens were scanned at a heating rate of 20°C/min.

Copolymer characterization by NMR spectroscopy

The average copolymer composition was determined by ^{1}H NMR analysis of copolymer samples dissolved in a mixture of deuterated benzene and tetrachloroethylene (5 wt%). The spectra were recorded at 70°C using an XL 100 Varian spectrometer (100 MHz for ^{1}H). For the characterization of monomer sequence distribution, ^{13}C NMR was used; Fourier transform (FT) ^{13}C NMR spectra were recorded at 70°C using a Bruker WP 80 spectrometer (20·1 MHz for ^{13}C). Copolymer samples were examined as 15·20 wt% solutions in a mixture of deuterated benzene and tetrachloroethylene (50/50 by volume), using tetramethysilane (TMS) as internal reference.

Colloidal properties

Particle size measurements

The particle sizes of the latexes were determined by light scattering from the dissymmetry measurements and by transmission electron microscope. Staining with uranyl acetate solutions was used in order to harden the soft copolymer particles for examination in the electron microscope. The weight average and number average diameters were calculated from the electron micrographs. The light scattering measurements were used to calculate weight average diameters.

Surface characterization of the latex particles

The latexes were diluted to 2·5% solids with deionized water and cleaned by the ion-exchange method using purified mixed-bed resins of Dowex 1 and Dowex 50W according to the method reported by Vanderhoff *et al.*[9] The ion-exchange process was repeated until a constant surface charge density was achieved. The conductometric titration method using 0·019 N sodium hydoxide solution was used to determine the amount of the acidic groups on the surface of the particles.

Determination of rates of coagulation

The kinetics of the early stages of coagulation was followed by recording the change in optical density with time, after the addition of the electrolyte solution to the 'cleaned' latex sample. All measurements were carried out in the infrared region at a wavelength of 1370 nm using the Cary-14 Spectrophotometer, according to the method described by Wu.[10] Latexes diluted in deionized water to about 0·4% by weight were used in all the coagulation experiments. Sodium and barium chloride and aluminum nitrate solutions were used as electrolytes. The initial slopes of the optical density–time curves were used to calculate the stability factor W for each electrolyte concentration. The critical coagulation concentration (c.c.c.) was determined from the electrolyte concentration at $W = 1$.

RESULTS AND DISCUSSION

Preparation–kinetic investigations

As summarized in Table I, two series of vinyl acetate-butyl acrylate copolymer latexes were prepared in a wide range of comonomer compositions and using two processes of emulsion polymerization: the conventional batch technique and the semi-continuous method. The emulsion polymerization of vinyl acetate-butyl acrylate comonomers is a complex process mainly due to wide differences in the reactivity ratios and the water solubility between the two monomers; as reported in Table II.

Hence, the method of monomer addition is expected to have a drastic effect on the compositional homogeneity of the different copolymer latexes. Since this parameter was found to be a key factor for interpreting many results in the present work, an attempt was made to gain a better knowledge of the emulsion polymerization of this system in order to simulate the kinetics of these different emulsion copolymerizations as a function of the comonomer composition in the feed and the polymerization process.

At first, kinetic studies were performed using gas chromatography analysis of the unreacted monomer mixture through the basic equipment described in Fig. 1 and ^{1}H NMR analysis of the copolymer composition versus conversion.

TABLE II
Basic parameters in vinyl acetate (VAc)–butyl acrylate (BuA) copolymers

(a) Reactivity ratios

Reactivity ratios R_{VAc}	Reactivity ratios R_{BuA}	*Experimental conditions*	*Ref.*
0	8	68°C (emulsion)	5
0·06	3·07	70°C (?)	11
0·0388	5·529	50°C (emulsion)	12
0·040	5·50	60°C (emulsion)	13

(b) Water solubility of monomers (20°C)

Vac = 2·5 g/100 g	BuA = 0·08 g/100 g

Knowledge of kinetic parameters such as initial feed compositions and, chiefly, reactivity ratios is essential in order to use the copolymer equation to calculate the monomer sequence distribution in radical-initiated copolymers. Because literature data about reactivity ratios are often questionable (they are found to be affected by the polymerization conditions such as temperature, solvent, process, etc.) the actual reactivity ratios for VAc–BuA systems were determined in the range of initial monomer feed compositions used for the preparation of the copolymer latexes. A non-linear least square method based

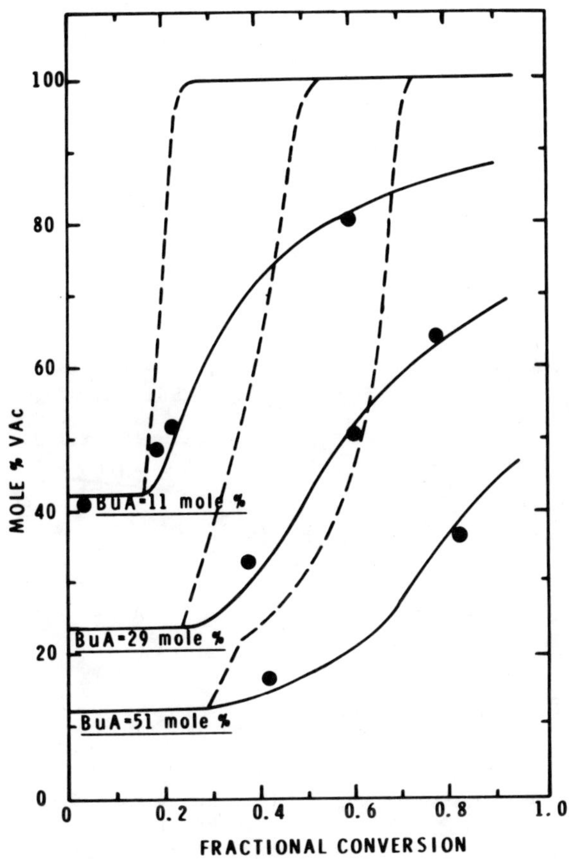

Fig. 2. Batch process simulation of copolymer composition versus conversion. Average (——); instantaneous (----) and experimental data from ^{1}H NMR analysis.

on Tidwell and Mortimer's idea[14] was then applied through a computer program using the copolymer composition-conversion data. The calculated reactivity ratio values[13] given in Table II are found to be very close to the recently published data related to the emulsion process.

The kinetics of the different emulsion copolymerization were then simulated using a single model fitted in the lab.[15] Basically, it needs experimental or available data (such as monomer feed composition, reactivity ratios of the system, water solubility of the two monomers, partition coefficient) and assumes that the polymerization takes place inside or near the surface of the polymer particle. Many interesting features of the copolymerization process can be derived from this model, especially the average and instantaneous copolymer compositions versus conversion—very useful data to characterize the extent of compositional heterogeneity. Figure 2 gives an example of simulation of batch emulsion copolymerizations and illustrates the effect of the butyl acrylate content on the average and instantaneous copolymer composition versus conversion. Good agreement is found between the experimental data and simulation curves. Moreover, it is remarkable to observe the important drift of the copolymer composition with the conversion, regardless of the butyl acrylate content. However, it may be observed that the higher the butyl acrylate content in the monomer feed, the larger the drift. On the

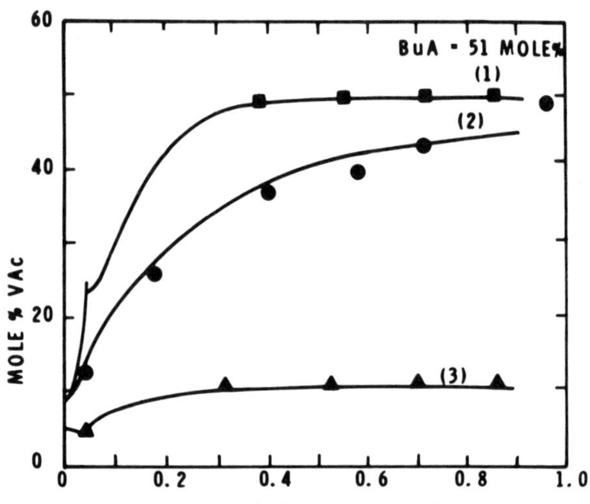

Fig. 3. Semi-continuous process simulation. Copolymer composition instantaneous (1), and average (2); unreacted VAc monomer (3). Experimental data from ^{1}H NMR analysis (■), (●), (▲).

other hand, these simulation curves also indicate the conversion at which pure poly(vinyl acetate) is produced because of the butyl acrylate depletion.

Figure 3 shows the simulation of the same features but for a semi-continuous polymerization in which the monomer addition rate was found to be practically equal to the rate of polymerization at the end of the batch period. The curves show that even using a relatively small initial charge, a notable drift appears which is slowly compensated after 50% conversion. This initial drift during the batch period must be correlated with the fact that due to the difference in water solubility a relatively large part of vinyl acetate monomer is in the aqueous phase whereas a large part of butyl acrylate is in the organic phase. This results in a change in the monomer feed composition compared to the theoretical value. This figure emphasizes the effect of the monomer addition rate on the homogeneity of the copolymer composition. As Fig. 4

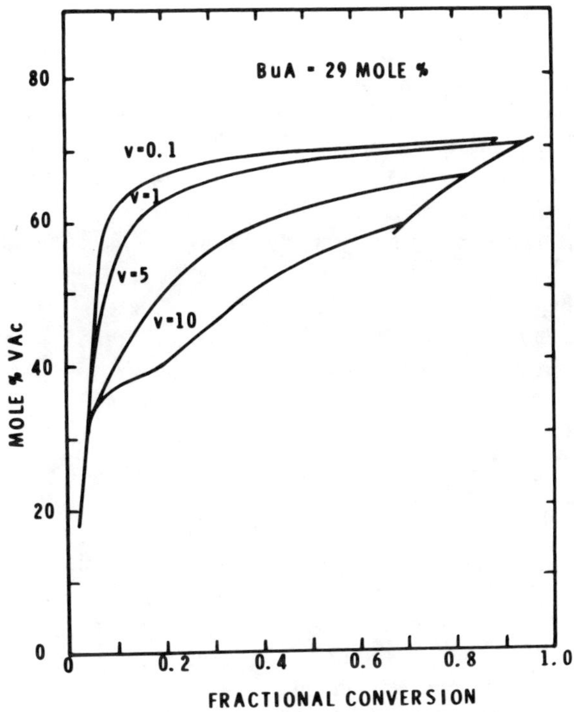

Fig. 4. Semi-continuous process simulation: effect of addition rate on the average copolymer composition versus conversion. v = Rate of monomer addition/rate of polymerization.

shows, it is obvious that when this addition rate becomes smaller the drift can be obviated and most of the copolymer produced has a constant composition. In return, a slow addition rate involves a small rate of polymerization and favors secondary reactions, such as branching reactions in vinyl acetate polymerizations (see next section).

The control of the homogeneity of the copolymer composition can be improved by using 'the composition controlled batch' method, a new process described in detail elsewhere[16] which was found to produce constant composition copolymer all over the polymerization period of the vinyl acetate–butyl acrylate.

Molecular properties of copolymers

Molecular weight and distribution (MWD)

MWD of homopolymers: Experimental molecular weights and distributions (MWD) are presented in Figs. 5 and 6 for PVAc and PBuA; they are found to be drastically affected by the emulsion polymerization process.

For PVAc, the batch process sample shows a broad distribution due to long chain branching at high conversion, as reported by Friis and Hamielec.[17] From Table III, the polydispersity ratio value (14·7) appears very close to that given by different authors. It was well demonstrated that this increase in the polydispersity ratio is mainly due to transfer to polymer, the grafting onto the main chain being more important than onto the acetyl side group.[18]

For PBuA, the distribution appears to be narrower than in PVAc, but the relatively high polydispersity ratio (11·5) still suggests some long chain branching occurring at high conversion onto the α-H of the main chain.

Differences in molecular weight may be related to the product of rate constants $k_p \times k_t^{-1/2}$ which is higher at 25°C for PBuA (1·2) than for PVAc (0·15).[19] Indeed the polymerization rate of BuA was found to be about eight times larger than that of PVAc.[5] This explanation is only qualitative because k_t is also diffusion controlled at high conversion.

For semi-continuous samples, chromatograms of the two homopolymers exhibit complex distributions. A clear bimodal distribution is observed for PVAc, the first peak in the low M_w region being related to a very broad distribution of polymer chains (polydispersity = 175); moreover, the small value obtained for M_n is an indication of the presence of oligomers. In addition to the main peak, a second very sharp peak is observed in the high molecular weight region, recently mentioned by Bataille *et al.*[20]

It may be assumed that this peak would also be caused by long chain branching structures produced all over the polymerization duration as a result

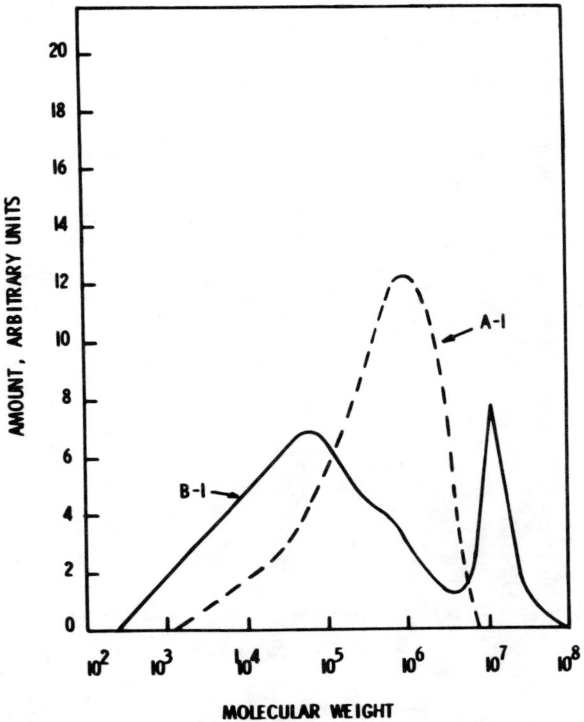

Fig. 5. Differential molecular weight distribution in poly(vinyl acetate) latex: batch process (A-1), and semi-continuous process (B-1).

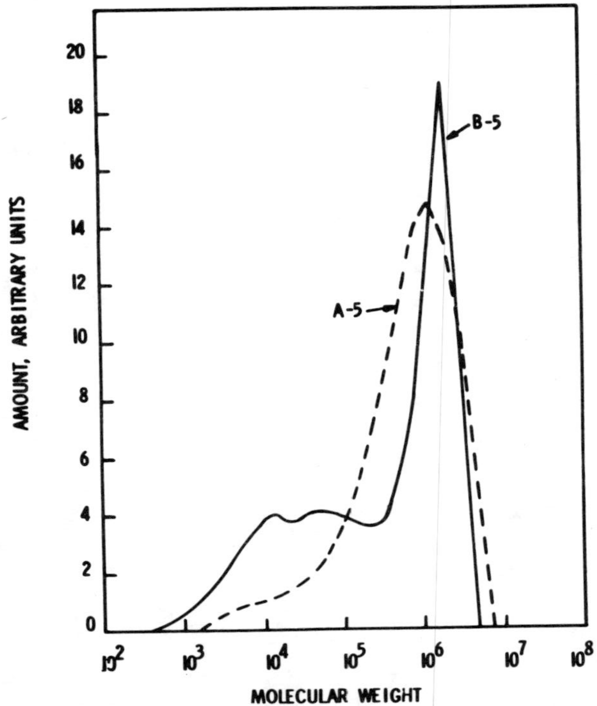

Fig. 6. Differential molecular weight distribution in poly(butyl acrylate) latex: batch process (A-5), and semi-continuous process (B-5).

of 'starved' conditions. Indeed, the monomer concentration always stays very low and constant at the particle surface; hence, transfer reactions to polymer can take place deep inside the particles, which is highly concentrated in polymer, and compete with the 'normal' propagation which occurs near the particle surface.

For PBuA, this second peak is found to be in higher relative proportion suggesting that transfer reaction to polymer also prevails in this homopolymer at high conversion.

MWD of copolymers: The same trend is confirmed for the copolymer samples as shown in Figs. 7 and 8, but some differences may be characteristic of copolymerization features.

For the batch copolymers, the MWD is only slightly dependent on the butyl acrylate content, the polydispersity ratios (Table III) being larger than in the homopolymers.

For the semi-continuous copolymers, low M_n values are observed regarding the first distribution, which is affected by this copolymer composition that may be related to the lower monomer concentration at the reaction locus in the semi-continuous process compared to the batch one,[2] as well as to the presence of a large amount of oligomers.

Concerning the second peak in the high molecular weight region, its intensity is found to be dependent on the copolymer composition since for one sample (when the BuA content is 29 mole %) only a shoulder is observed which cannot be separated from the first distribution. For this particular composition, the molecular weight distribution (see last column in Table III), which is also dependent on the copolymer composition, is found to be the most narrow ($M_w/M_n = 8{\cdot}68$).

Dynamic mechanical properties

Vinyl acetate and butyl acrylate homopolymers: The emulsion polymerization process (batch and semi-continuous) was found to have a considerable effect on the glass transition temperature and the dynamic mechanical behavior of both homopolymers and copolymers of vinyl acetate and butyl acrylate.

The glass transition temperature (*Tg*) of all the samples, as obtained through DMS and DSC, is summarized in Table IV. Both DMS and DSC show that poly(vinyl acetate) produced through the batch process (sample A-1) (Fig. 9) has a *Tg* higher by 6°C than its counterpart (sample B-1) produced through the semi-continuous process. This indicates that the mode of emulsion polymerization has an effect on the structure of poly(vinyl acetate).

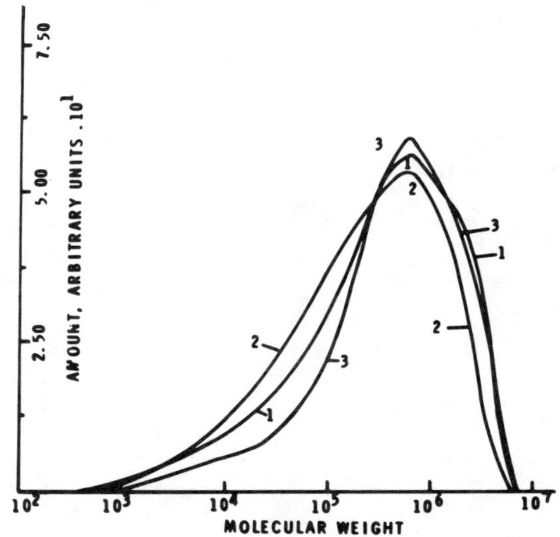

Fig. 7. Differential molecular weight distribution in batch VAc–BuA copolymer latexes. Identification of curves is given as mole % of BuA: 11 (1), 29 (2), and 51 (3).

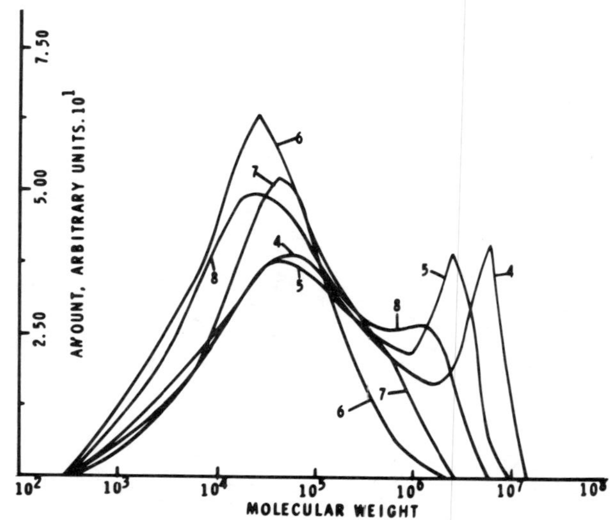

Fig. 8. Differential molecular weight distribution in semi-continuous VAc–BuA copolymer latexes. Identification of curves is given as mole % of BuA: 11 (4), 18 (5), 29 (6), 40·5 (7), and 51 (8).

TABLE III
Molecular weight characterization by gel permeation chromatography of VAc–BuA copolymer latexes

Latex composition BuA/VAc mole ratio	*Number average molecular weight* M_n ($\times 10^4$)	*Weight average molecular weight* M_w ($\times 10^5$)	M_w/M_n
Batch			
0/100	6·527	9·623	14·74
11/89	4·748	9·950	20·96
29/71	4·748	7·727	16·27
51/49	5·196	10·290	19·81
100/0	10·760	12·360	11·50
Semi-continuous			
0/100	0·721	12·650	175·43
11/89	1·827	15·460	84·62
18/82	1·609	8·850	55·01
29/71	0·955	0·830	8·68
40·5/59·5	1·113	1·690	15·18
51/49	1·855	4·594	24·76
100/0	1·556	9·636	61·93

TABLE IV
Glass transition temperatures of vinyl acetate–butyl acrylate copolymer films

Composition VAc/BuA mole ratio	*Tg*† (°C)				
	Batch (A-series)			*Semi-continuous (B-series)*	
	DSC	*DMS*	ΔT‡	*DSC*	*DMS*
100/0	35	60	—	27	54
89/11	−15 & 21	0 & 40	40	21	38
71/29	−33 & 17	−17 & 28	45	4	18
49/51	−45 & 31	−32 & 43	75	−20	5
0/100	−53	−40	—	(−73) & −53	(−58) & −40

† Absolute value of *Tg* depends on the heating rate and the frequency at which the sinusoidal strain is applied to the sample, and also on the mode of testing.
‡ ΔT = difference in *Tg*s of the individual peaks and is a measure of heterogeneity.

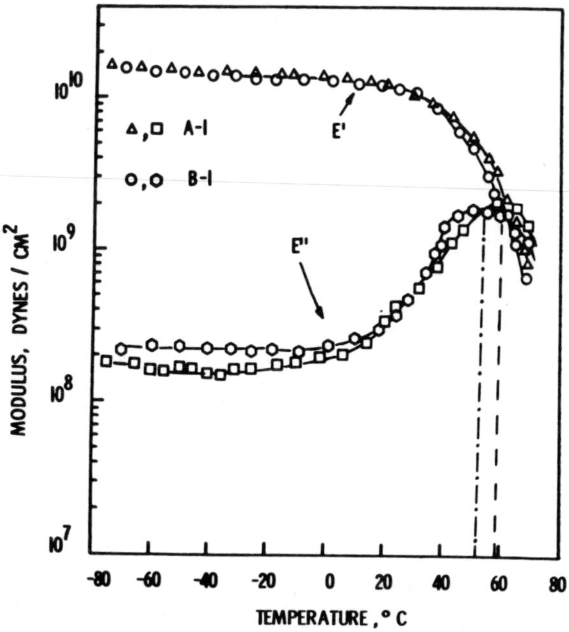

Fig. 9. Dynamic mechanical spectra of poly(vinyl acetate) latex films produced through the batch process (A-1) and the semi-continuous process (B-1). E' = storage modulus and E'' = loss modulus.

The glass transition temperature is known to depend, among other factors, on the degree of branching and the molecular weight distribution. It is well known that branching occurs in the vinyl acetate polymerization at high conversion.[21] Even though the poly(vinyl acetate) samples A-1 and B-1 were polymerized to 100% conversion, it is expected that the degree of branching in the batch will be different from that in the semi-continuously prepared latex. In the semi-continuous latex B-1, the polymerization was carried out under starved conditions, i.e. conversion was close to 100% all the time, thus it is expected that the degree of branching will be quite extensive. On the other hand, in the case of the batch latex sample (A-1), branching can only take place during the last stage of the polymerization, e.g. above 80% conversion. Thus, the extent of branching is expected to be less in the batch latex than in the semi-continuous latex. Based on this speculation alone, one would expect the *Tg* of the batch latex to be less than that of the semi-continuous latex. However, the contrary was found to be true experimentally.

On the other hand, the GPC-molecular weight was found to be quite different for these two latexes (see Fig. 5). The batch latex showed only one band with a number average molecular weight ($6{\cdot}52 \times 10^4$), whereas the semi-continuous latex showed two bands; the first one was quite broad, 10^2–10^6, with a number average molecular weight of ($0{\cdot}66 \times 10^4$) and a second narrow band with a number average molecular weight of ($119{\cdot}4 \times 10^4$). It is quite possible that the presence of the larger amounts of the low molecular weight polymer in the semi-continuous latex acts as a plasticizer, and counteracts the effect of the excessive branching; thus causing the *Tg* of the semi-continuous latex to be less than the batch latex.

On the other hand, no difference was observed in the *Tg* of poly(butyl acrylate) (samples A-5 and B-5) prepared through the batch or the semi-continuous process. However, a small shoulder was observed at -58°C in the loss modulus spectrum (Fig. 10) of sample B-5 (prepared through the semi-continuous process), which indicates the presence of some low molecular weight polymer, consistent with the GPC results shown in Fig. 6.

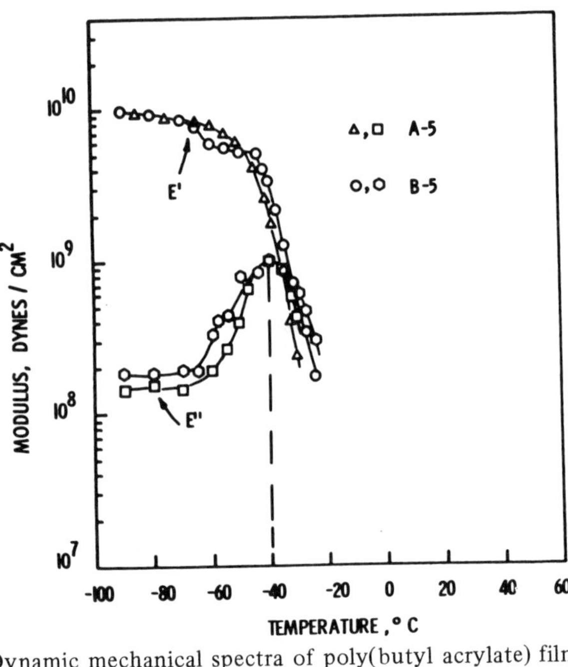

Fig. 10. Dynamic mechanical spectra of poly(butyl acrylate) films produced through the batch process (A-5) and the semi-continuous process (B-5). E' = storage modulus and E'' = loss modulus.

Furthermore, the peaks observed in the loss modulus spectra were broader for samples prepared by the semi-continuous process indicating that these have a broader distribution of molecular weight. Indeed, GPC results had shown that the samples prepared through the semi-continuous process have a broader distribution of molecular weight. This is because in batch polymerization, all the monomer is added in at the beginning, which results in a relatively short particle-generation time as compared to the semi-continuous process (where the monomer feeding rate is smaller than the polymerization rate). The low monomer concentration at the locus of polymerization in the semi-continuous process can cause the formation of low molecular weight polymer, relative to the batch polymerization process, as suggested earlier by Krackeler and Naidus.[2] Thus, the longer particle generation time in the semi-continuous process results in a broad distribution of molecular weight and in some cases in a bimodal distribution.

Vinyl acetate–butyl acrylate copolymers: The loss modulus spectra of all the films prepared from copolymer latexes produced through the batch process showed two peaks corresponding to the glass transition temperature (Tg) of the butyl-acrylate-rich and vinyl-acetate-rich copolymers. In comparison, the films prepared from the semi-continuous polymerized copolymer latexes showed a single peak, as shown for example in Fig. 11. These results indicate that the latexes prepared through the batch process have an inhomogeneous composition having two components; one rich in butyl acrylate, and the other rich in vinyl acetate. On the other hand, the single peak observed in the latexes produced through the semi-continuous process indicates a more homogeneous composition, corresponding to the monomer mixture. The butyl-acrylate-rich core observed in the latter case, through electron microscopy, is too small to be distinguished by dynamic mechanical spectroscopy. The loss modulus spectra of the batch polymerized copolymers also indicates that the distance between the two transition peaks increases with an increase in the butyl acrylate composition (ΔT in Table IV). This indicates that the heterogeneity in the copolymers increases with an increase in the butyl acrylate content of the monomer mixture. These results are in good agreement with the electron microscopy[22] and the stability studies of these latexes discussed elsewhere.[23]

Monomer sequence distribution

In order to obtain quantitative information on the chemical composition homogeneity in these copolymers, a microstructure study was investigated using ^{13}C nuclear magnetic resonance (^{13}C NMR) spectroscopy.[13] This tech-

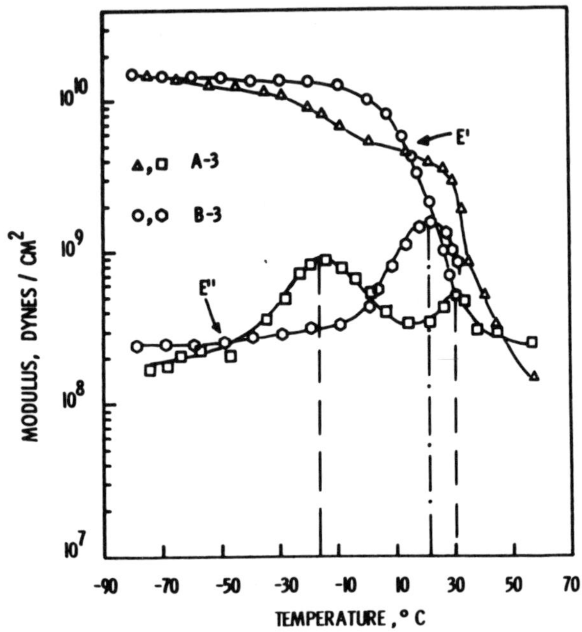

Fig. 11. Dynamic mechanical spectra of (63 : 37) vinyl acetate–butyl acrylate copolymer latex films produced through the batch process (A-3) and semi-continuous process (B-3).

nique has been used to examine differences in the monomer sequence distributions between the two series of vinyl acetate (A)-butyl acrylate (B) copolymers.

After the assignment of all the observed ^{13}C lines in the obtained spectra of the homopolymers and copolymers, the methine and methylene carbon resonance regions were used for the calculation of the VAc centered triads fractions and diads proportions.

In Fig. 12 is shown a typical example corresponding to the expanded (A) methine region spectra of two copolymers having the same composition (49% mole A) but produced through the two polymerization processes. Each main peak was attributed to the different A-centered triads, respectively BAB, AAB and AAA. It should be mentioned that the methine carbon resonances of A units display further splitting due to the tacticity effect. Qualitatively, noticeable variations may be observed in the intensity of the three main peaks R_1, R_2, R_3 as a function of the polymerization process.

Regarding the (B) methine and methylene carbon resonance part of the

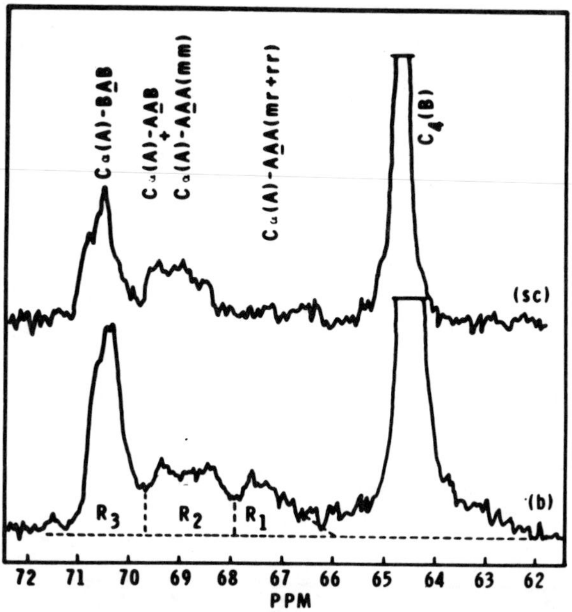

Fig. 12. ^{13}C resonance (20·1 MHz) of the methine carbons of vinyl acetate (A)–*n*-butyl acrylate (B) copolymers with 49% mole acetate obtained in batch (b—bottom) and semi-continuous (sc—top) processes.

same spectra (Fig. 13), more drastic discrepancies are disclosed—higher proportions of long butyl acrylate sequences (BBB triads and BB diads) in the batch copolymers compared to the semi-continuous ones which contain more alternated structures (such as ABA triads and AB diads).

From this ^{13}C NMR analysis, all A-centered triad and diad proportions were quantitatively determined for both series of copolymers. The determined values, given in Tables V and VI, are also compared to the calculated ones deduced from first order Markov statistics and taking into account the degree of conversion; fairly good agreement is observed within 10%. Because of strong overlapping, the B-centered triad proportions cannot be accurately determined and are not reported.

Comparison of the A sequence distribution between the two series of copolymers, clearly confirms the previous observations arising from the spectra. Copolymers prepared through the semi-continuous process have a more homogeneous composition and indeed the A-centered triad and diad fractions appear to be very close to the calculated values. For the batch

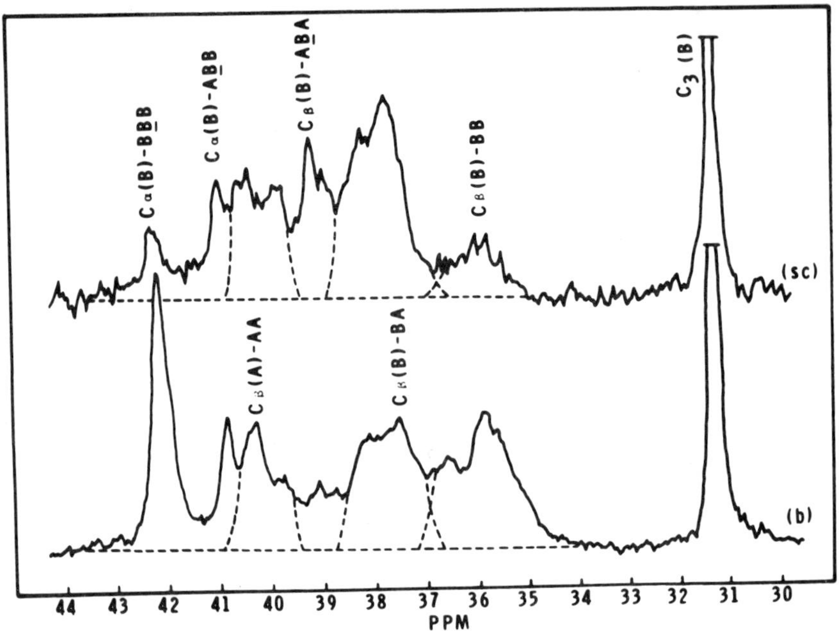

Fig. 13. ^{13}C resonances (20·1 MHz) of the methylene carbons of vinyl acetate units and the methine and methylene carbons of *n*-butyl acrylate units of vinyl acetate (A)–*n*-butyl acrylate (B) copolymers with 49% mole vinyl acetate obtained in batch (b—bottom) and semi-continuous (sc—top) processes.

copolymers, a very heterogeneous composition is found, ranging from butyl-acrylate-rich fractions to A-rich copolymers and pure poly(vinyl acetate). For example, concerning the 49/51 [A]/[B] composition, the batch copolymer contains a higher amount of A-centered AAA triad parts (0·30) than the semi-continuous one (0·04), but a lower amount of A-centered AAB triad fraction (0·23) as compared to 0·425 in the latter. Moreover the BB diad proportion is found to be three times higher in the batch copolymer. This trend is confirmed for the other copolymer compositions.

The results also suggest that batch polymerization gives rise to the formation of poly(vinyl acetate) because of the rapid consumption of B in the monomer feed. Since it was found that a quite good agreement exists between experimental and calculated A-centered triad and diad fractions, the same computer program may be used to determine the variation of [B] in the monomer feed and of the instantaneous composition of the copolymer versus conversion. Figure 14 clearly illustrates these variations for the three monomer

TABLE V
Experimental (^{13}C NMR) and calculated vinyl acetate centered triad fractions of vinyl acetate (A)–butyl acrylate (B) copolymer samples. I: Batch. II: Semi-continuous

	Samples					
	I	II	I	II	I	II
Overall composition (mole ratio)	[A]/[B] = 49/51		[A]/[B] = 71/29		[A]/[B] = 89/11	
BAB experimental	0·47	0·53	0·22	0·18	0·09	—
BAB calculated†	0·52	0·48	0·29	0·13	0·08	0·01
AAB experimental	0·23	0·42	0·12	0·43	0·11	0·21
AAB calculated†	0·16	0·43	0·16	0·46	0·13	0·20
AAA experimental	0·30	0·04	0·66	0·39	0·79	0·79
AAA calculated†	0·32	0·09	0·55	0·41	0·78	0·78

† Calculated values from reactivity ratios $r_A = 0{\cdot}04$, $r_B = 5{\cdot}50$

TABLE VI
Experimental (^{13}C NMR) and calculated diad fractions of vinyl acetate (A)–butyl acrylate (B) copolymer samples. I: Batch. II: Semi-continuous

	Samples					
	I	II	I	II	I	II
Overall composition (mole ratio)	[A]/[B] = 49/51		[A]/[B] = 71/29		[A]/[B] = 89/11	
BB experimental	0·39	0·13	0·16	0·08	—	—
BB calculated†	0·35	0·17	0·14	0·04	0·03	—
AB experimental	0·35	0·61	0·30	0·49	0·16	0·19
AB calculated†	0·31	0·67	0·29	0·51	0·17	0·21
AA experimental	0·26	0·25	0·53	0·44	0·83	0·80
AA calculated†	0·33	0·15	0·56	0·45	0·80	0·78

† Calculated values from reactivity ratios $r_A = 0{\cdot}04$, $r_B = 5{\cdot}50$

feeds which allows estimation of the conversion at which poly(vinyl acetate) begins to be produced. It also confirms the qualitative observation derived from the dynamic mechanical spectroscopy study regarding the increase in the magnitude of the compositional heterogeneity with an increasing propor-

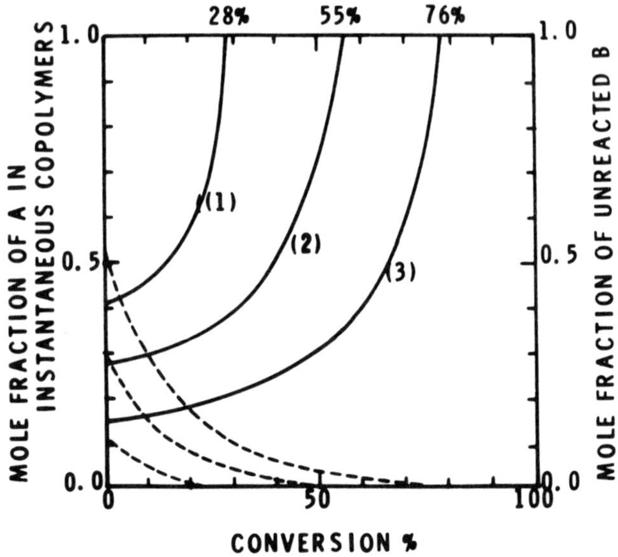

Fig. 14. Instantaneous copolymer compositions (——) and unreacted *n*-butyl acrylate mole fraction (----) versus conversion with 11 (1); 29 (2); 51 (3) mole % *n*-butyl acrylate in the initial monomer feed. Poly(vinyl acetate) homopolymer begins to appear at 28, 55 and 76% conversion for (1), (2) and (3) respectively.

tion of butyl acrylate in the monomer feed. It may be interesting to compare Fig. 14 with that related to the viscoelastic behavior (Fig. 15).

At first, the drift in the instantaneous copolymer composition drastically increases in the copolymer sample (respectively from 1 to 3) which results in an enlargement of the first peak corresponding to the *Tg* of the B-rich copolymer compositions. Secondly, a good correlation is also found between the amount of poly(vinyl acetate) produced through the batch process and the relative intensity of the second peak corresponding to this homopolymer ($Tg \cong 40°C$); this peak becomes larger and larger as compared to the amount of PVAc produced, which increases from 24, to 45, to 72% in the different samples (3), (2) and (1) respectively.

Colloidal properties of latex

Particle size and size distribution

The particle size results are summarized in Table VII. A good agreement was found between the light scattering results and the weight average dia-

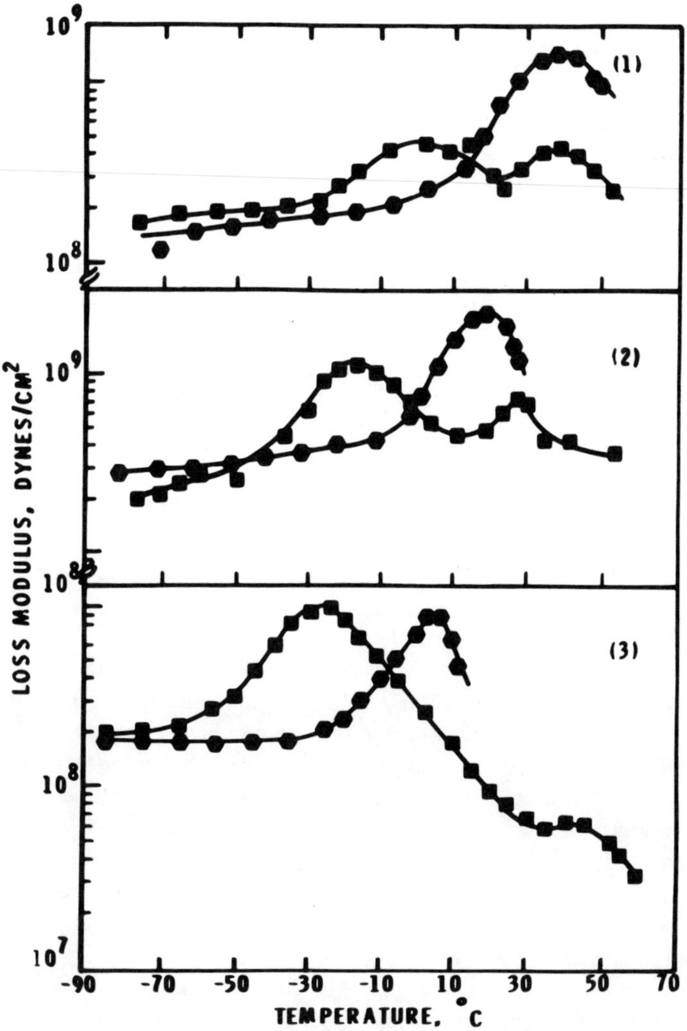

Fig. 15. Dynamic mechanical spectra (loss modulus) of vinyl acetate–butyl acrylate copolymers prepared by batch (■) and semi-continuous (⬣) emulsion polymerization. Copolymers with 11 (1); 29 (2) and 51 (3) mole % butyl acrylate.

meters determined by the electron microscopy method. In general, the average particle diameter for the semi-continuous latexes was found to be smaller than the batch-latexes with the exception of the poly(vinyl acetate) where the reverse was true.

TABLE VII
Particle size of latexes

Latex composition BuA/VAc mole/ratio	*Light scattering* D_w *(Å)*	*Electron microscopy*		*Poly-dispersity* D_w/D_n
		D_n *(Å)*	D_w *(Å)*	
Batch				
0/100	1160	880	1188	1·35
11/89	1270	1060	1170	1·10
29/71	1330	900	970	1·07
51/49	1170	1090	1232	1·136
100/0	960	—	—	—
Semi-continuous				
0/100	1620	1650	1720	1·043
11/89	1000	930	1000	1·081
18/82	945	745	845	1·134
29/71	690	520	930	1·786
40·5/59·5	760	730	773	1·056
51/49	790	—	—	2
100/0	585	—	—	large

This difference in the particle size of the copolymer latexes could be explained on the basis of the ideas of Krackeler and Naidus.[2] The duration of particle formation is much longer and the number of particles is higher in the semi-continuous process compared to that in the batch process. Thus, the particle size is expected to be smaller and the distribution to be broader in the semi-continuous latexes. On the other hand, the larger size and narrower distribution found in the case of the semi-continuous poly(vinyl acetate) latex compared to the batch-latex could be attributed to Dunn's ideas[24] regarding the coalescence of the small size particles with the larger particles during the polymerization.

The average particle diameter of the batch copolymer latexes was found to be independent of the copolymer composition. On the other hand, the average particle diameter of the semi-continuous copolymer-latexes was found to decrease, and the poly-dispersity to increase, with increasing butyl acrylate content in the copolymer. This difference in particle size dependence on the copolymer composition of the latexes prepared by the two processes could be explained in terms of the difference in the locus of initiation in the two processes, and the difference in the aqueous phase solubility and the reactivity ratios of the two monomers. In the batch process, initiation is known to take place predominantly in micelles.[25] Thus in this case, the particles are formed over a short period of time—early on during the initial stages of polymerization. The particle growth takes place mainly by diffusion of monomer and polymerization within the particles swollen with monomer. Due to the difference in reactivity ratios between the two monomers, the final particles are comprised of poly(butyl acrylate)-rich core and poly(vinyl acetate)-rich shell (which has been demonstrated elsewhere).[22, 26] Thus, the average particle size is not expected to depend to any extent on the copolymer composition in the batch latexes. On the other hand, in the case of the semi-continuous process, the particle nucleation is assumed to take place predominantly in the aqueous phase. Thus, the particles are formed slowly, over a long period of time during the addition of the comonomer mixture. The particle growth takes place by coalescence of smaller size particles or low molecular weight oligomers with larger size particles, as well as by diffusion of monomer and polymerization within the particle. Due to the starved monomer condition during the course of the polymerization (rate of monomer addition was kept lower than the rate of polymerization), the formed particles will be homogeneous with respect to copolymer composition. In the case when the comonomer mixture contains a high content of the relatively more hydrophobic, butyl acrylate, monomer, the surface charge density of the newly formed small particles will be high due to the good packing of the emulsifier molecules on the surface. Thus, the stability of these small size particles against coalescence with larger size particles increases, which results in a latex system with small average size and broader distribution. As the butyl acrylate content in the comonomer composition becomes less, the small latex particles—now rich in vinyl acetate—will present a relatively more hydrophilic surface with the result of less emulsifier adsorption per unit surface area. This can lead to less stability of the small size particles with the result of the coalescence among themselves, or coalesce with an existing large size particle, to form larger size particles. This in turn should lead to an increase in the average particle diameter and narrowing the particle size distribution, with the increase in the vinyl acetate content in the copolymer latex.

Surface characterization

The end groups on the surface of the latex particles could come from one or more of the following sources. (i) *Initiator fragments*—each polymer molecule should contain two end groups. In a persulfate initiated emulsion polymerization, the end groups are usually sulfate, hydroxyl or carboxylic groups. (ii) *Hydrolysis of the poly(vinyl acetate) chains*—which can result in the formation of hydroxyl groups. (iii) *Hydrolysis of the poly(butyl acrylate) chains*—which can result in the formation of carboxylic acid groups.

The concentrations of the surface hydroxyl groups were not determined in the present work; however, their presence will be evident in the stability studies as shown in the following sections.

The characterization results with respect to the surface sulfate, carboxyl end groups and the sum of the two, i.e. the total, are summarized in Table VIII. All latexes, except the batch poly(vinyl acetate), were found to have both sulfate and carboxyl groups on the surface of their particles.

TABLE VIII
Surface characterization results of VAc–BuA copolymer latexes

Latex composition BuA/VAc mole ratio	Concentration of end groups (μeq/g polymer)			Number of acid end groups per polymer molecule	Surface charge density (μC/cm)	
	Strong acid	Weak acid	Total		Strong acid	Weak acid
Batch						
0/100	2·36	—	2·36	0·15	0·52	—
11/89	2·22	11·13	13·35	0·63	0·54	2·70
29/71	3·10	12·64	15·74	0·75	0·79	3·21
51/49	5·76	14·18	19·94	1·04	1·29	3·17
100/0	2·05	8·80	10·85	1·17	0·32	1·35
Semi-continuous						
0/100	10·90	12·50	23·30	0·17	3·4	3·9
11/89	9·45	11·85	21·30	0·39	1·78	2·24
18/82	10·45	11·30	21·75	0·35	1·38	1·49
29/71	14·31	19·30	33·30	0·34	1·37	1·85
40·5/59·5	12·60	18·80	31·40	0·35	1·67	2·50
51/49	10·45	24·80	35·25	0·65		
100/0	8·60	16·3	24·90	0·39		

The surface concentration of the acid end groups is a complex function of the average molecular weight and the molecular weight distribution of the polymer; the degree of hydrolysis of the sulfate end groups to hydroxyls and oxidation to carboxyl during the course of polymerization; and the degree of 'burial' of the end groups inside the particles. It is expected then, that the polymerization process should affect the surface concentration of the acid end groups with the same magnitude as it affects the above factors.

The surface concentrations of the sulfate, the carboxyl and the sum of the two, which are given in Table VIII in units of μ-equivalent per gram of polymer, are also shown in Fig. 16. The results show that the total surface concentration of the acid groups in the semi-continuous latex particles is much higher than in the batch latex particles. This difference could be

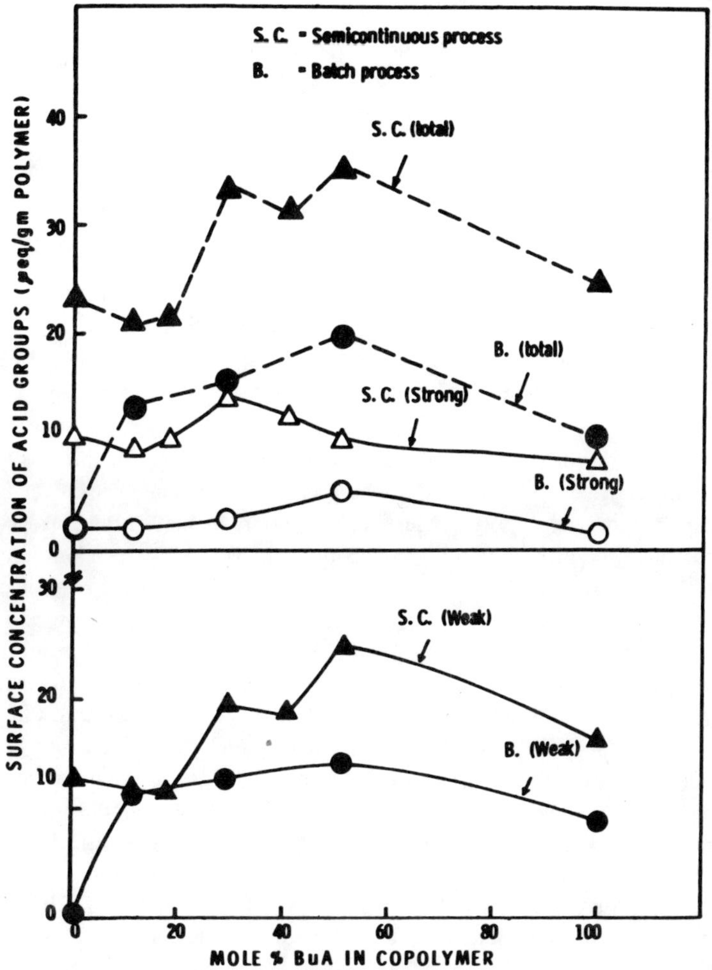

Fig. 16. Surface concentrations of the sulfate, the carboxyl and the sum of the two (in μ-equivalent per gram of polymer) as a function of copolymer composition and polymerization process.

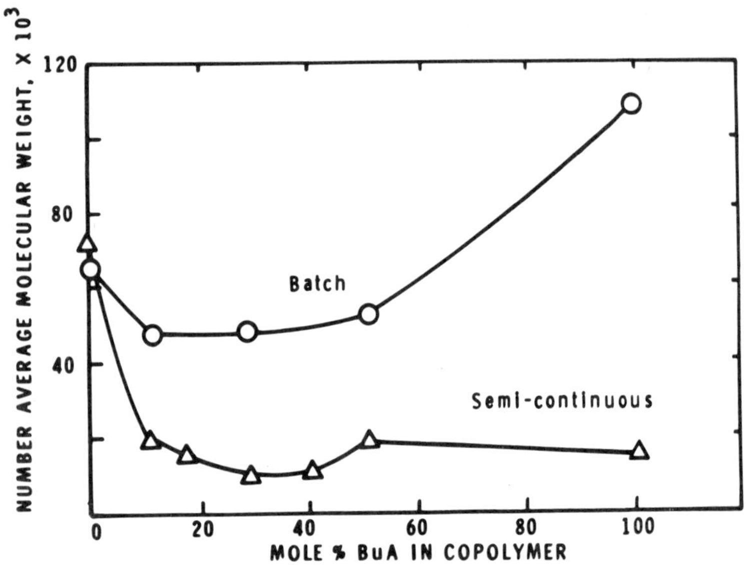

Fig. 17. Number average molecular weight ($\times 10^3$) versus copolymer composition: batch process (○); semi-continuous process (△).

explained on the basis of the difference in molecular weight of the two types of polymers. Figure 17 shows that the number average molecular weight of the semi-continuous copolymer latexes is lower than that of the corresponding batch copolymer latexes. Recognizing that each polymer chain should theoretically have two end groups, the polymer with the lower molecular weight is expected to have a higher concentration of end groups per unit weight of the polymers, which is shown to be the case for the semi-continuous latexes. The number of the surface acid end groups per polymer molecule in the batch and the semi-continuous latex particles, which is given in the fifth column of Table VIII, was calculated from the total surface acid end groups per one gram of polymer given in the same table, and the number average molecular weight given in Table III. In both latexes the number of surface acid end groups per polymer molecule is less than the theoretical value of two. This is taken as an indication of 'burial' of the acid end groups inside the particles and/or hydrolysis of the end groups to hydroxyl groups. The results also show that the number of surface acid end groups per polymer molecule in the semi-continuous latexes is less than in the batch latexes. Given the fact that the time of polymerization in the semi-continuous process was shorter (5·5 hours) than in the batch process (24 hours), this eliminates the hydrolysis as being the reason for such a difference between the two types of latexes. Consequently, it seems that the lower number of acid end groups per polymer molecule in the semi-continuous latex could be due to the higher degree of 'burial' inside the particles. No attempt was made in this study to determine the concentration of the 'buried' end groups inside the latex particles. However, some indication of their presence will be evident in the section on the ageing effect on the surface characterization of the latexes.

Another feature of the data given in Fig. 16 is that the surface concentration of the total acid end groups in the semi-continuous copolymer latexes is more dependent on the copolymer composition than in the batch latexes. This difference is better explained by considering the dependence of the concentration of the sulfate and carboxyl end groups separately on the copolymer composition, and the polymerization process.

The concentration of the surface sulfate end groups in the semi-continuous copolymer latexes was found to be higher than that of the batch latexes (see Fig. 16). In addition to the difference in the number average molecular weight between the two types of latexes, the effect of the polymerization process on the hydrolysis of the sulfate end groups could be another reason for such a difference. The longer polymerization time of 24 hours at 70°C in the batch process compared to 5·5 hours in the semi-continuous process could cause more hydrolysis and oxidation of the sulfate end groups in the batch latexes relative to the semi-continuous latexes. The surface sulfate end groups' concentration in the semi-continuous latexes was found to increase to a maximum with increasing butyl acrylate content in the copolymer up to 29 mole %, then decrease thereafter with further increase in the butyl acrylate content. On the other hand the surface sulfate end groups' concentration in the batch latexes was found to be almost independent of the copolymer composition. These differences seem to be related to the effect of the copolymer composition on the number average molecular weight (see Fig. 17), and the molecular weight distribution of the copolymer (see Table III). The results shown in Fig. 16 and Table III show that, in general, the molecular weight is much lower, and the distribution is much 'broader', in the semi-continuous copolymer latexes compared to the batch latexes. On the other hand the average molecular weight and the molecular weight distribution in the batch latexes are almost independent of the copolymer composition.

The surface carboxyl end group concentration was found to be higher in the semi-continuous copolymer latexes with butyl acrylate content above 19 mole %, than in the batch latexes. Also the surface carboxyl end groups' concentration was found to vary with the copolymer composition in the semi-continuous latexes, but to be almost independent of the copolymer composition in the batch latexes. The difference in the homogeneity of the copolymer composition within the particle (see earlier) can account for such differences. The presence of poly(vinyl acetate)-rich shell in all the batch latexes can explain the unchanged concentration of the surface carboxyl groups with the change in the overall composition, since in this case the surface carboxyl groups can result mainly from the hydrolysis and oxidation of the surface sulfate groups. On the other hand, with a much more homogeneous copolymer composition in the semi-continuous latexes, and with the possibility of hydrolysis of some of the poly(butyl acrylate) chains to give carboxyl groups at the surface, one can account for the higher surface carboxyl content and their dependence on the copolymer composition.

Effect of ageing and hydrolysis on the surface properties of latexes

The effect of ageing and acid hydrolysis on the surface properties of the latexes are given in Table IX. The ageing experiments were carried out at room temperature by letting the 'cleaned' latexes sit on the shelf in their bottles for a period of six months, followed by ion-exchange cleaning and characterization. The acid hydrolysis was done by treating the 'cleaned' latex with $1{\cdot}5 \times 10^{-3}$ M solution of hydrochloric acid at 50°C for five hours followed by ion-exchange cleaning and characterization. Ageing and hydrolysis can

TABLE IX
Ageing effect on surface characterization

Latex composition BuA/VAc mole ratio	*eq/g Polymer* Before ageing		After ageing for six months		% Change in surface end groups	
	Strong	*Weak*	*Strong*	*Weak*	*Strong*	*Weak*
Batch						
51/49	5·76	14·18	5·83	17·0	+ 1·2	+ 19·9
Semi-continuous						
0/100	10·90	12·50	7·85	16·60	+ 28·0	+ 32·8
11/89	9·45	11·85	8·51	28·63	−10·0	+ 126·4
51/49	10·45	24·80	14·20	46·60	+ 35·9	+ 87·9
Semi-continuous (Hydrolyzed in 0·0015 N HCl medium, at 50°C; 5 h)						
11/89	9·45	11·85	9·40	39·00	−0·5	+ 229·1
51/49	10·45	24·80	21·00	104·00	+ 100·9	+ 319·3

result in one or more of the following transformations: (i) vinyl acetate to vinyl alcohol, (ii) butyl acrylate to carboxyl groups, and (iii) sulfate groups to hydroxyl and/or carboxyl groups. The rate of hydrolysis of each species in the copolymer should depend on the homogeneity of the copolymer composition within the particle and the monomer sequence in the copolymer.

The autocatalytic acid hydrolysis of the poly(vinyl acetate) at the surface of the latex particles is known to cause the formation of polyvinyl alcohol. The acid hydrolysis of vinyl acetate takes place due to the electrodonating affinity of the acetyl group ($-COCH_3$) towards the proton (H^+-ions); thus weakening the $RO-COCH_3$ bond with the result of formation of acetic acid and hydroxyl groups. This hydrolysis is autocatalytic in nature due to the adjacent group effect of the vinyl alcohol on the unreacted vinyl acetate units. This type of hydrolysis is expected to depend on the homogeneity of the copolymer. The calculated average length of the vinyl acetate sequence (Fig. 18) was found to drop drastically from a very high value, with the

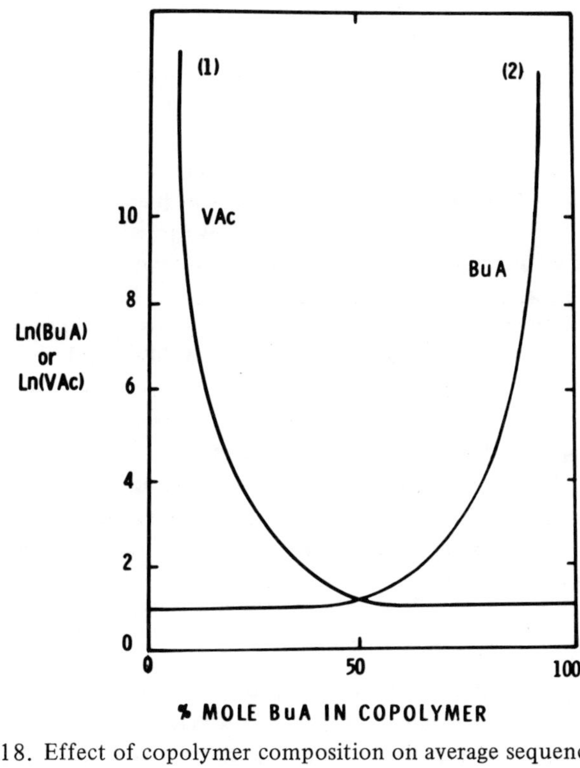

Fig. 18. Effect of copolymer composition on average sequence length.

increase in butyl acrylate content in the copolymer, to attain a value close to 1 at a 50 : 50 molar ratio of the two monomers. The rate and magnitude of the autocatalytic hydrolysis of vinyl acetate groups is expected to decrease rapidly with the decrease in the vinyl acetate sequence length in the copolymer. However, due to the difficulty in determining the surface concentration of the hydroxyl groups, this change was not checked experimentally. However, these changes in the surface hydroxyl concentration were found to affect the colloidal stability of these latexes.

The results in Table IX show that the method of monomer addition and the copolymer composition have a profound effect on the results of the hydrolysis of the surface acid groups. The surface concentrations of the sulfate and carboxyl groups change only slightly after hydrolysis of the batch 51/49 butyl acrylate-vinyl acetate copolymer latex and the semi-continuous poly(vinyl acetate) latex. This, again, is expected on the basis of the similarity in the chemical nature for the two latexes, i.e. the presence of poly(vinyl acetate) on the surface of the particles.

The hydrolysis results in the semi-continuous copolymer latexes showing a remarkable change in the surface concentration of the sulfate and carboxyl groups. The magnitude of this change is found to depend on the copolymer composition, but no clear trend can be suggested. Indeed, if the percentage change in the surface carboxyl groups after ageing for six months is due to hydrolysis of the butyl acrylate group one expects to see a lower change in the 11/89, butyl acrylate-vinyl acetate copolymer latex compared with the change in the 51/49 copolymer latex; the reverse is true. However, the expected change in weak acid content was found upon subjecting these two latexes to severe acid hydrolysis (see Table IX).

The increase in the surface concentration of the strong acid groups, which was found upon ageing or hydrolysis of the 51/49 butyl acrylate-vinyl acetate copolymer latex sample prepared by the semi-continuous method, is difficult to explain. However, the possibility exists that some of the low molecular weight polymer chains ending with sulfate groups, which were buried inside the particles, are released from the interior to the surface of the particles. This could be aided by the increase in the surface carboxyl groups (due to the hydrolysis of the butyl acrylate) which can cause swelling and plasticization of the particle during the neutralization with alkali.

Colloidal stability

The critical coagulation concentration (c.c.c.) in moles/liter of the added mono- and divalent electrolytes to the 'cleaned' latexes are given in Table X. The c.c.c. values for the semi-continuous latexes were found to decrease

TABLE X
Electrolyte stability results

Latex composition BuA/VAc mole/ratio	*c.c.c. (millimole/liter)*		$\frac{(c.c.c.)\ BaCl_2}{(c.c.c.)\ NaCl}$
	NaCl	*$BaCl_2$*	
Semi-continuous			
0/100	257	51·9	0·202
11/89	72·3	18·6	0·257
18/82	75·8	19·0	0·251
29/71	123·0	29·0	0·236
40·5/59·5	141·2	29·0	0·205
51/49	423·0	76·0	0·180
Batch			
11/89		28·8	
29/71		20·0	
51/49	80·6	22·40	0·278

sharply with the introduction of a low content of butyl acrylate in the copolymer, but then it slowly increases with further increase in the butyl acrylate content. This seems to follow the same trend of the change in the surface charge density (due to strong and weak acid groups) as a function of the copolymer composition as shown in Table VIII. This could be taken as an indication that the electrostatic stabilization mechanism is active in the colloidal stability of this class of latexes. The ratio of the c.c.c. for the divalent to the monovalent electrolyte does not appear to be proportional to the inverse sixth power of the cation valency as expected from the Schulz–Hardy rule. Instead the ratio is much closer to the inverse second power of the cation valency, such an effect is predicted by the DLVO theory for the stability of particles with low surface charge density.[27]

The c.c.c. values of the divalent electrolyte for the batch latexes were found not to be affected by the surface charge density nor by the copolymer composition.

The stability against electrolyte of these latexes was found to be strongly affected by ageing and hydrolysis. A comparison between the c.c.c. of the mono- and divalent electrolytes before and after ageing and hydrolysis of three of the semi-continuous latexes is given in Table XI. Also given in the same table is the percentage change in the amount of electrolyte due to ageing or hydrolysis, which should be compared with the corresponding percentage change in the surface end groups as a result of the hydrolysis (see

TABLE XI

Effect of hydrolysis on colloidal stability of BuA–VAc copolymer latexes prepared by the semi-continuous method

Latex composition BuA/VAc mole/ratio	*c.c.c. (millimole/liter)*				*% Change in amount of electrolyte to reach c.c.c.*	
	Before		*After*			
	NaCl	*$BaCl_2$*	*NaCl*	*$BaCl_2$*	*NaCl*	*$BaCl_2$*
0/100 (Ageing 6 months)	259·0	51·9	2000·0	224	+ 672·2	+ 331·6
11/89 (Hydrolyzed, HCl, 5 h)	72·4	18·6	165·0	45·8	+ 127·9	+ 146·2
51/49 (Hydrolyzed, HCl, 5 h)	423·0	76·0	403·0	109·0	−4·7	+ 43·4

Table VIII). These results show that the percentage change in the amount of the electrolyte required to reach the c.c.c. does not correspond to the percentage change in the surface density which took place as a result of the ageing or the hydrolysis.

This is taken as an indication of the increased importance of the polyvinyl alcohol at the surface layer as a result of the autocatalytic hydrolysis of the poly(vinyl acetate) sequences, which can cause the steric stabilization which is more important than the electrostatic stabilization due to the sulfate or the carboxyl surface groups.

Therefore, it seems obvious that the colloidal stability of these copolymer latexes is also dependent on the morphology and structure of the surface layer which are known to be affected by the homogeneity of the copolymer composition, the degree of branching, the degree of hydrolysis and the flexibility of the polymer molecules.

The effect of hydrolysis of these latexes is even more pronounced in the results given in Table XII for the stability against trivalent electrolyte, after ageing for a period longer than one year.

The stability is found to be almost independent of the copolymer composition for the batch latex whereas a strong dependence is showed for the semi-continuous latex. The highest value of the c.c.c. was found for the 'aged' poly(vinyl acetate) latex, in which, the particles are probably fully covered

TABLE XII
Colloidal stability against trivalent electrolyte of 'cleaned' BuA–VAc copolymer latexes after ageing for one year

Latex composition BuA/VAc mole/ratio	*c.c.c. (millimole/liter)*	
	Batch	*Semi-continuous*
0/100	20	2000
11/89	20	46
18/82	—	6·0
29/71	26	6·5
40·5/59·5	—	9·5
51/49	13	0·58
100/0	17	5·8

with a layer of polyvinyl alcohol as a result of the hydrolysis; the c.c.c. was found to decrease very sharply with increasing butyl acrylate content, the lowest value being formed for the 51/49 butyl acrylate–vinyl acetate copolymer latex. This follows the expected trend in the decrease in the autocatalytic hydrolysis of the poly(vinyl acetate) to polyvinyl alcohol, due to the decrease in the sequence length of the vinyl acetate in the copolymer. The fact that the c.c.c. for the aged poly(butyl acrylate) is much less than that for the 'aged' poly(vinyl acetate) latex suggests the lesser efficiency of the carboxyl groups (produced by hydrolysis of the butyl acrylate groups upon ageing) as a steric stabilizer compared to the hydroxyl groups.

The much higher value of the c.c.c. of the trivalent electrolyte required for the semi-continuous poly(vinyl acetate) latex compared to the batch latex may be explained on the basis of the difference in the morphology of the surface layer in the two latexes. The surface layer in the semi-continuous latex could be visualized as being more 'fuzzy' than the batch latex, with the hydrolyzed polyvinyl alcohol chains extending farther out from the surface of the particles into the aqueous phase; thus causing the highest stability. This difference in particle morphology could conceivably be the result of different degrees of branching which is known to take place during the polymerization process at high conversion.[17] In the semi-continuous process, the polymerization was carried out under 'starved' monomer condition, i.e. conversion was close to 100% all the time. Consequently, it is expected that the degree of branching will be quite extensive. On the other hand in the batch latex, the branching can only take place during the last stages of the polymerization, i.e. above 80% conversion. Thus, the branching is expected to be less extensive in the batch latex than in the semi-continuous latex. This coupled with the hydrolysis of the exposed poly(vinyl acetate) groups to polyvinyl alcohol can cause the difference in 'fuzzyness' of the surface in the two types of latexes.

CONCLUSIONS

The comonomer composition and the method of monomer addition were found to have a remarkable effect on the bulk properties as well as on the surface and colloidal properties of vinyl acetate–butyl acrylate copolymer latexes.

At first, owing to a kinetic study, it was shown that the butyl acrylate monomer was rapidly consumed, as expected; which results in large copolymer composition deviations versus conversion in batch copolymerization. The semi-continuous process was found to better control the copolymer composition, but the drift cannot be avoided in the initial batch period.

Batch copolymer latexes have a narrower molecular weight distribution, whereas semi-continuous latexes have a broader molecular weight distribution and in many copolymer compositions bimodal distributions are obtained, an indication of severe branching.

Dynamic mechanical properties of latex films showed that batch copolymers are heterogeneous with respect to the copolymer composition whereas the semi-continuous copolymers are more homogeneous.

A quantitative determination of the monomer sequence distribution in the copolymers confirms that the batch copolymerization process gives heterogeneous copolymers having long homosequences, while with the semi-continuous process, the obtained copolymers have a better statistical distribution of monomers.

Particle size distribution and the surface characteristics of the copolymer latexes were found to vary with the copolymer composition and the emulsion polymerization process. The ageing of latexes was shown to affect the surface properties to different degrees depending on the degree of homogeneity of the copolymer composition and the monomer sequence length in the copolymer.

The colloidal stability against electrolytes of batch polymerized latexes is much less dependent on copolymer composition than latexes prepared by semi-continuous method. The effect of ageing showed the importance of the steric stabilization due to the formation of polyvinyl alcohol chains; it was found to be strongly related to the degree of homogeneity of the copolymer composition with the particle.

All these results confirm that in batch-polymerized latexes, the particles have a core–shell morphology with the core rich in the more reactive component (butyl acrylate), and the shell rich in the less reactive (vinyl acetate); on the other hand, the semi-continuous latex particles have a more homogeneous copolymer composition per particle, representative of the composition of the two monomers. These differences in particle morphology were clearly shown in an electron microscope study using a staining technique (where the butyl-acrylate-rich regions are preferentially statined).[26] Moreover these differences in morphology were found to affect the film formation of these copolymers (as described in another paper[22]); where the further gradual colescence particularly was much slower in the batch polymerized latexes as compared to the semi-continuous polymerized latexes.

ACKNOWLEDGMENT

The financial support of The National Science Foundation to C. Pichot under the US–France Exchange Program is gratefully acknowledged.

REFERENCES

1. H. Gerrens, *J. Polym. Sci., C,* **27**, 77 (1969).
2. J. J. Krackeler and H. Naidus, *J. Polym. Sci., C,* **27**, 207 (1969).
3. G. W. Poehlein, *Advances in emulsion polymerization and latex technology,* 10th Annual Short Course, June 1979, Emulsion Polymers Institue, Lehigh University, PA.
4. J. Snuparek and F. Krska, *J. Appl. Polym. Sci.,* **20**, 1753 (1976).
5. K. Chujo, Y. Harada, S. Tokuhara and K. Tanaka, *J. Polym. Sci., C,* **27**, 321 (1969).
6. L. Rios, C. Pichot and J. Guillot, *Makromol. Chem.,* **181**, 677 (1980).
7. E. B. Bradford and J. W. Vanderhoff, *J. Macromol. Chem.,* **1**, 355 (1966).
8. E. B. Bradford and J. W. Vanderhoff, *J. Macromol. Sci. Phys.,* **B6**, 671 (1972).
9. J. W. Vanderhoff, H. J. Van den Hul, R. J. M. Tausk and J. Th. G. Overbeek, in *Clean Surfaces,* G. Goldfinger (ed.), Dekker, New York, 1970, p. 15.
10. W. C. Wu, *Preparation and Surface Characterization of Latexes as Model Colloids,* Ph.D. Dissertation, Lehigh University, 1977.
11. A. F. Nikolaev, L. P. Vishnets Kaya, O. H. Gromova, M. M. Grigoreva and M. S. Kleshcheva, *Vysokomolekul Soedin, Ser. A,* **11**, 2418 (1969).
12. I. F. Avetisyan, K. A. Pospelova, K. E. Onikul and P. I. Zubov, *Labokrasochnye Materialy i ikh Primenemie,* (2), 13 (1964); *Chem. Abstr.,* **61**, 9677a (1964).
13. C. Pichot, M. F. Llauro and Q. T. Pham, Submitted to *Europ. Polym. J.*
14. R. W. Tidwell and G. H. Mortimer, *J. Polym. Sci., A1,* **3**, 369 (1969).

15. L. Rids, C. Pichot and J. Guillot, To be published.
16. A. Guyot, J. Guillot, C. Pichot and L. Rids, *Organic Coatings and Plastics Preprints,* **43**, 629 (1980).
17. N. Friis and A. E. Hamielec, *J. Appl. Polym. Sci.,* **19**, 97 (1975).
18. S. I. Nozakura, Y. Morishima and S. Murahashi, *J. Polym. Sci.–Polym. Chem. Ed.,* **10**, 2853 (1972).
19. J. Brandrup and E. H. Immegrut (eds.), *Polymer Handbook,* Wiley, New York, 1975, p. II.45.
20. P. Bataille, B. T. Van and Q. B. Pham, *J. Appl. Sci.,* **22**, 321 (1979).
21. W. W. Graessley, R. D. Harting and W. C. Uy, *J. Polym. Sci., A2,* **3**, 1919 (1969).
22. S. C. Misra, C. Pichot, M. S. El-Aasser and J. W. Vanderhoff, *J. Polym. Sci.,* accepted for publication.
23. C. Pichot, T. Makgawinata, M. S. El-Aasser and J. W. Vanderhoff, Abstracts, 176th ACS Meeting, Miami Beach, Florida, Sept. 10–15, 1978; *Polymer Preprints,* **19**, 870 (1979).
24. A. S. Dunn and L. C. H. Chong, *Br. Polym. J.,* **2**, 49 (1970).
25. M. Numura, M. Harada, W. Eguchi and S. Nagata, *ACS Symposium Series,* No. 24, 1976.
26. S. C. Misra, C. Pichot, M. S. El-Aasser and J. W. Vanderhoff, *J. Polym. Sci., B–Polym. Letters Ed.,* **17**, 567 (1979).
27. C. Silebi, W. C. Wu, M. S. El-Aasser and J. W. Vanderhoff, Paper presented at the 176th ACS Meeting, Colloid and Surface Chemistry Division, Miami Beach, Florida, Sept. 11–14, 1978.

13

Some Studies on Vinyl Acrylic Latex – Surfactant Interactions

B. R. VIJAYENDRAN, T. BONE and C. GAJRIA

Celanese Plastics and Specialties Company, Jeffersontown, Kentucky, USA

SUMMARY

The effects of vinyl acetate monomer polarity and its propensity to undergo chain transfer reactions on surfactant interactions in poly(vinyl acetate) homopolymer and poly(vinyl acetate–acrylate) copolymer latexes are investigated. First, adsorption of several surfactants commony used in emulsion polymerization at a fairly hydrophilic (polar) poly(vinyl acetate) latex surface are compared to those at a more hydrophobic (less polar) polystyrene latex surface. Optimum HLB for the adsorption of non-ionic surfactants at a hydrophilic poly(vinyl acetate) and at a hydrophobic polystyrene latex surface are determined and compared with available literature data. Surfactant adsorption results presented in this section are qualitatively explained by considering the relative hydrophobic/hydrophilic nature of surfactant and polarity of latex surface.

Second, adsorption of non-ionic surfactants at a vinyl acrylic copolymer latex/water interface is correlated to the measured polarity of the latex surface, in agreement with previous work on the adsorption of sodium lauryl sulfate at various polymer/water interfaces. The idea of using contact angle measurements on dried latex films and surfactant adsorption to characterize the polarity and composition of copolymer latex surfaces is proposed.

Third, some of the specific interactions of anionic surfactants in vinyl acrylic latexes are considered. Effect of surfactant structure, polymer composition, presence of adsorbed nonionic surfactants at the latex surface, and other factors, on the shapes of adsorption isotherms and on solubilization leading to latex thickening are considered.

Fourth, effects of nonionic and anionic surfactants on glass transition temperature, minimum film formation temperature and film formation process in vinyl acrylic latexes are described. Latex film morphology and the further gradual coalescence phenomena in vinyl acrylic latexes are studied and the results critically compared with the available literature for the film formation process in styrene–butadiene and all acrylic latexes.

Fifth, literature on the grafting of nonionic surfactants and polymeric stabilizers in poly(vinyl acetate) latexes is reviewed. Application of the serum replacement technique to study grafting of nonionic surfactants in vinyl acrylic latexes is noted. The need for systematic work to study grafting of nonionic stabilizers in vinyl acrylic latexes is also indicated.

The emulsion polymerization behavior of vinyl acetate is distinguished by two factors. One is the solubility of vinyl acetate in water of about 2·8%. The other factor is the propensity of vinyl acetate radicals to undergo chain transfer reactions. It is generally accepted that these two features of vinyl acetate significantly alter its emulsion polymerization kinetics compared to those of styrene, and the Smith–Ewart mechanism is not applicable to the emulsion polymerization of vinyl acetate.[1]

Surfactants play a central role in emulsion polymerization. For instance, stability of monomer emulsion droplet, reaction site and kinetics, particle size and stability of latex, properties and performance of final latex are all influenced to some degree or other by the nature of surfactant interactions in latexes. Such interactions have been studied to a limited extent in vinyl acetate and vinyl acetate–acrylate based latex. The present paper attempts to investigate in some detail the effects of one of the two unique features of vinyl acetate, namely its polarity—a measure of water solubility—on surfactant interactions in vinyl acetate and vinyl acrylic latex systems. Effects of chain transfer reactions in vinyl acetate polymerization leading to grafting of nonionic surfactants are reviewed to some extent. The effects of surfactants on reaction kinetics and mechanisms form another broad and important area of investigation in emulsion polymerization and they are not considered in this paper. More specifically, the following aspects of surfactant interactions are considered.

(a) Comparative adsorption behavior of surfactants at model polystyrene and poly(vinyl acetate) latex surfaces.
(b) Effect of polymer polarity of vinyl acrylic latex on nonionic surfactant adsorption.
(c) Specific interactions in the adsorption of anionic surfactants on vinyl acrylic latex.

(d) Effect of surfactants on glass transition temperature (*Tg*), minimum film formation temperature (MFFT) and film coalescence of vinyl acrylic latex.
(e) Grafting of nonionic surfactants.

EXPERIMENTAL SECTION

Materials: Commercial grade vinyl acetate (VAc) and butyl acrylate (BA) from Celanese Chemical Company were used. Reagent grade potassium persulfate was used as the initiator. Sodium lauryl sulfate (NaLS) from BDH Chemicals after purification[2] was used as the emulsifier in the polymerization.

Surfactants used in the adsorption studies of polystyrene (PS) and poly(vinyl acetate) (PVAc) are shown in Table I.

Surfactants used in the studies of vinyl acrylic latexes are Igepal CO-630 (10 mole ethylene oxide adduct of nonyl phenol), Alipal EP-110 and EP-120

TABLE I
Surfactant properties

Surfactant	*Critical micelle concentration*	
	Experimental	*Literature*
Ionics:		
Sodium lauryl sulfate (NaLS)	8·0 mM	7·8–8·2 mM
Sodium oleate	1·4 mM	1·2–1·5 mM
Hexadecyl trimethyl Ammonium bromide ($C_{16}TABr$)	0·4 mM	0·32 mM
Nonionics		
Poly(oxyethylene) nonyl phenyl ether–$C_9H_{19}C_6H_4(OCH_2 \cdot CH_2)_m \cdot OH$		
Igepal CO-710[a] $m = 10$	0·05 g/liter	0·05–0·06 g/liter
Igepal CO-850 $m = 20$	0·14 g/liter	0·14–0·20 g/liter
Igepal CO-880 $m = 30$	0·24 g/liter	0·28–0·48 g/liter
Igepal CO-970 $m = 50$	0·62 g/liter	0·68 g/liter

[a] A product of GAF corporation
m = No. of ethylene oxide units

from GAF Corporation, Aerosol A-102 from American Cyanamid and NaLS from BDH Chemicals. According to the manufacturer's literature, Alipal EP-110 and EP-120 are ammonium salts of a sulfated nonyl phenoxy poly-(ethyleneoxy) ethanol of molecular weights 708 and 1640 respectively. Aerosol A-102 is an anionic/nonionic type surfactant of the composition disodium ethoxylated alcohol (C_{10}–C_{12}) half esters of sulfo-succinic acid of molecular weight of about 1800.

Emulsion polymerization

A typical recipe is given in Table II. Emulsion polymerization was carried out at 60°C under a nitrogen atmosphere using a batch process. Theoretical solids content in all the formulations was 25%, and generally the conversions were better than 98%. Poly(vinyl acetate) (PVAc) and polystyrene (PS) homopolymers and two poly(vinyl acetate-butyl acrylate) copolymers having VAc/BA composition of 85/15 and 70/30 were prepared according to the above procedure.

TABLE II
Latex preparation

Monomer	25 g
Potassium persulfate	0·123 g
Sodium lauryl sulfate[a]	0·868 g
Distilled water	75 g

[a] Purified BDH product
Polymerizations were carried out at 60°C to 100% conversion

Ion exchange and dialysis of latexes

The latexes were 'cleaned' by the ion-exchange method of Van den Hull and Vanderhoff[3] and dialysis. In general, the latexes were stable after the cleaning process. In some cases, the ion-exchange resin was found to discolor after the treatment, indicating some interaction, perhaps hydrolysis of the latex, between latex and ion-exchange resin.

Particle-size determination

Particle-sizes of the 'cleaned' latexes were determined using transmission electron microscopy after freeze-drying the samples and counting the particles with a Quantimet image analyzer. The number average particle diameter (d_n)

of the homopolymers, the 85/15 VAc/BA and 70/30 VAc/BA latexes were found to be 0·057 μm, 0·062 μm and 0·073 μm, respectively. The polydispersity (d_w/d_n) of the three samples was 1·09, 1·05 and 1·06, respectively.

ADSORPTION STUDIES

Surfactant adsorption data presented under Results and Discussion, Section (a) at PS latex/water interface were obtained using the soap titration method of Maron *et al.*[4] The isotherms were derived from the soap titration data following the procedure of Paxton.[5] The single point data at the critical micelle concentration of surfactants at the PVAc latex/water interface were obtained following the method described below.

For the data presented in Sections (b) and (c), the following procedure was used.

Appropriate amounts of 'cleaned' latex and surfactants were mixed, equilibrated for 24 hours and the excess surfactant in serum analyzed after separation by centrifugation. Igepal CO-630 and Alipal surfactants were analyzed by UV absorption at 275 nm. NaLS and Aerosol A-102 were analyzed by colorimetric titration with cetyltrimethylammonium chloride in the presence of methylene blue and a chloroform layer, similar to the method of Epton.[6] Some nonionic surfactants and solubilized serum components were found to have a positive interference in the colorimetric analysis of ionic surfactants.

Contact angle measurements and polarity of latex films

Advancing contact angle measurements were made at 25°C using a Rame Hart contact angle goniometer. Latexes 'cleaned' by the above procedure were drawn on a glass plate and dried at 40–50°C. The average of eight measurements was taken as the contact angle. Water and methylene iodide were used as test liquids. Polarity of the dried latex films were estimated according to the method of Kaelble.[7]

RESULTS AND DISCUSSION

(a) Surfactant adsorption at model polystyrene and poly(vinyl acetate) latex surfaces

Adsorption isotherms of surfactants at polystyrene latex/water interface were determined by the soap titration method following the procedure of Paxton.[5]

The method consists of doing soap titrations at three latex concentrations (p) as shown in Fig. 1 for the PS–NaLS system. Surfactant concentrations C_1, C_2 and C_3 corresponding to the three latex levels at several surface tension values (γ) up to critical micelle concentration (CMC) are determined. Then plots of C versus p for the various γ values are made as shown in Fig. 2. Slope (Γ) and intercept (C_0) of the various lines give the amount of surfactant adsorbed in moles/g polymer and the bulk surfactant concentration in solution respectively. A plot of the adsorption isotherm Γ versus C_0, is shown in Fig. 3. Γ is converted to the corresponding area per molecule data (A) by using the expression

$$A = \frac{9{\cdot}961 \times 10^{-3}}{\Gamma \times \rho \times D} \tag{1}$$

where A = area per surfactant molecule, nm^2; D = diameter of latex, nm; ρ = density of polymer in g/cc.

A plot of A versus $1/C_0$ for the PS–NaLS system is shown in Fig. 4. The values of area per molecule of NaLS at polystyrene/water interface at CMC(A_m) and limiting concentration (A_0) are found to be 0·45 nm^2 and 0·28 nm^2, in good agreement with literature values.[8, 9]

Similar adsorption data (Γ versus C_0 and A versus $1/C_0$) for the other two ionic (Na oleate and C_{16} TABr) surfactants in the polystyrene system are

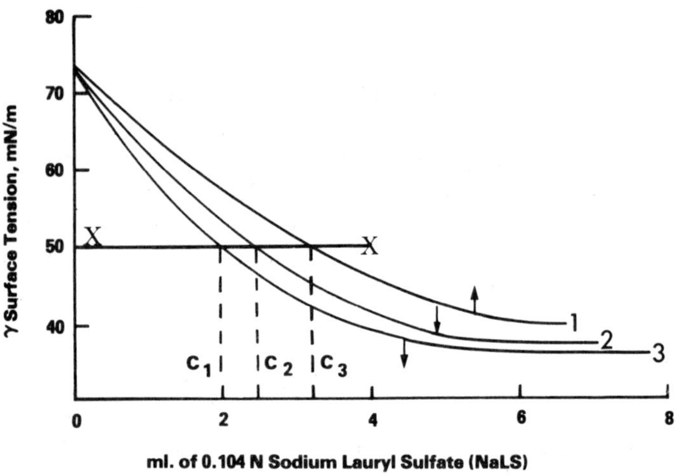

Fig. 1. Soap titration of ion-exchanged polystyrene latex with NaLS surfactant. Plots of γ versus ml of NaLS solution at three polymer concentrations. (1) = 1·0%, (2) = 0·5% and (3) = 0·1% latex solids.

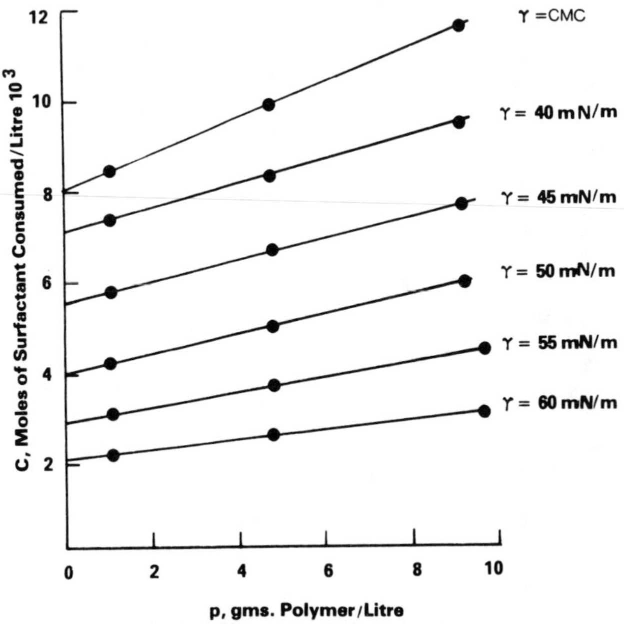

Fig. 2. Plots of C versus p at various surface tensions for PS–NaLS system.

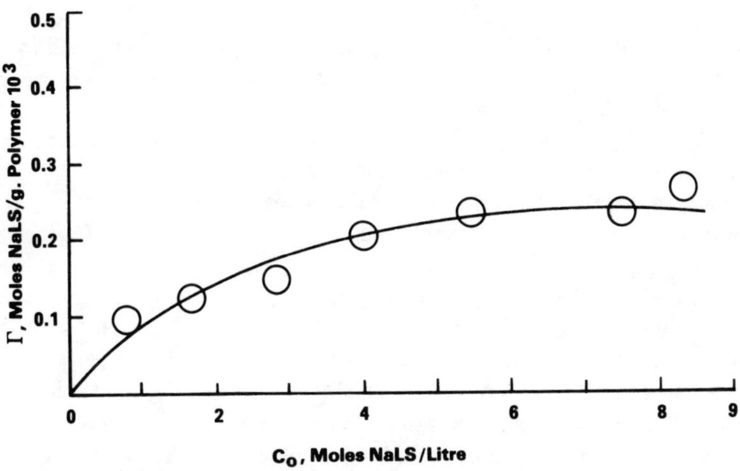

Fig. 3. Adsorption isotherm of NaLS at PS latex/water interface.

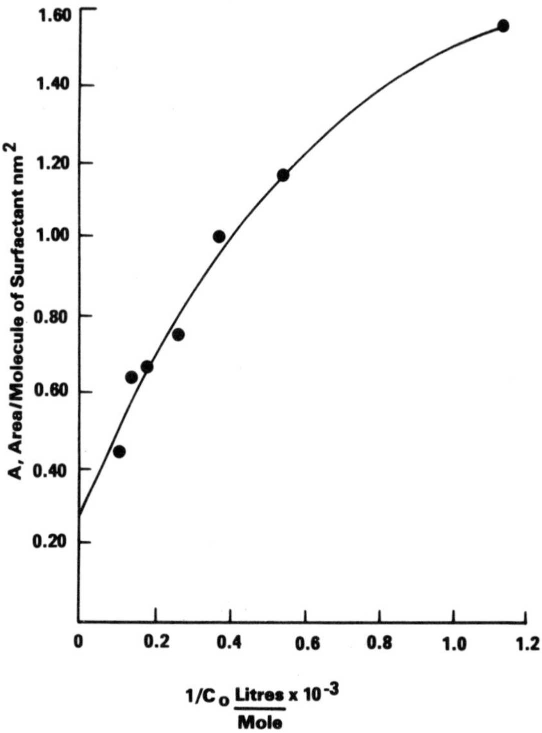

Fig. 4. Area per molecule versus $1/C_0$ for PS–NaLS system.

are shown in Figs. 5 and 6, respectively. A_m and A_0 values were found to be 0·30 nm^2 and 0·23 nm^2 for CTABr, respectively. These values are in excellent agreement with literature data.[10,11]

Similar adsorption data for a few of the nonionic surfactants studied at a polystyrene latex surface are shown in Figs. 7 and 8.

The above described soap titration method does not work satisfactorily for the poly(vinyl acetate) latex system since the end-points at CMC are not sharp and other reactions such as solubilization (discussed in detail in Section (b)) interfere with the measurement. Hence, for the poly(vinyl acetate) latex system, only single point adsorption value (Γ_m) corresponding to the CMC of surfactant was determined. A_m values corresponding to the area per surfactant molecule at CMC at a PVAc latex/water interface were calculated according to eqn. (1). Since complete adsorption isotherms were not obtained with the PVAc system, limiting A_0 values could not be determined. However, Paxton[5] has shown that A_0 value is characteristic of the surfactant molecule and is

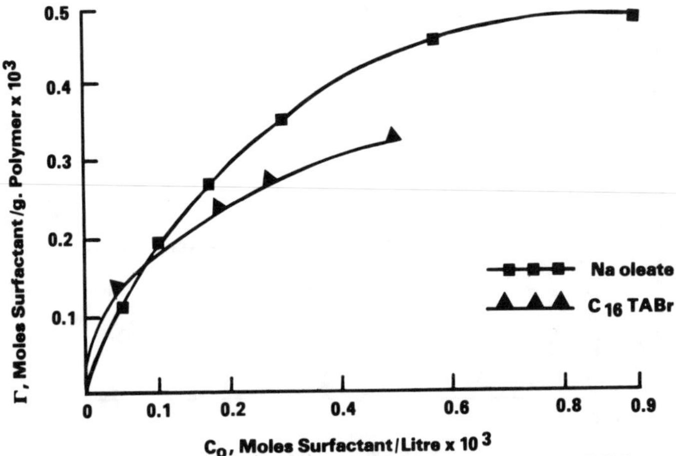

Fig. 5. Adsorption isotherm of Na oleate and C_{16} TABr at PS latex/water interface.

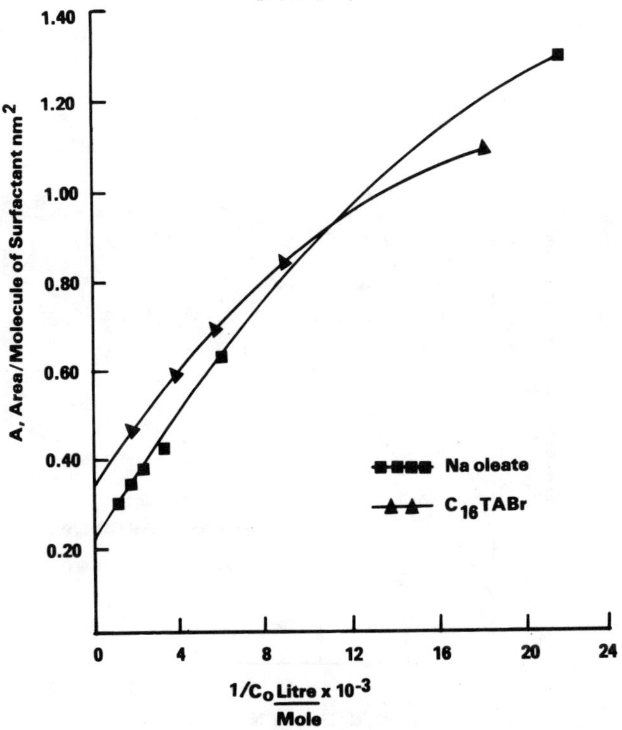

Fig. 6. Area per molecule versus $1/C_0$ for PS–Na oleate and PS–C_{16} TABr systems.

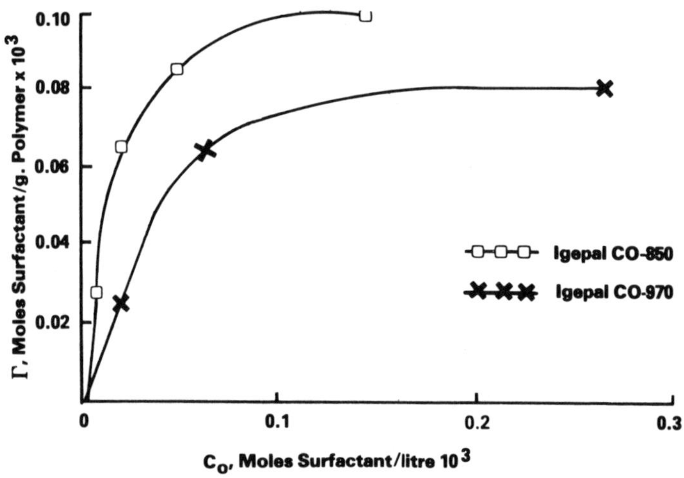

Fig. 7. Adsorption isotherm of nonionic surfactants at PS latex/water interface.

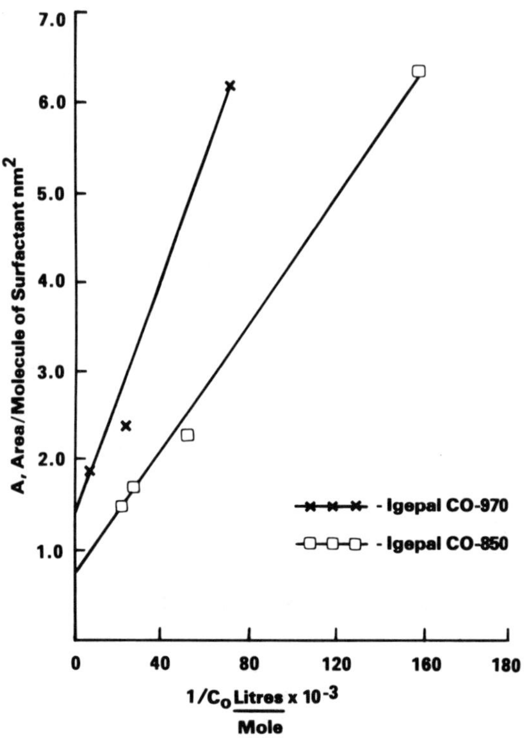

Fig. 8. Area per molecule versus $1/C_0$ for PS–nonionic surfactant systems.

independent of the latex surface. So A_0 values for the various surfactants determined above with the PS latex were used in the PVAc system also.

In order to compare the relative adsorption of different surfactants on latex surfaces of varying adsorption characteristics (polarity), it is convenient to use the ratio A_0/A_m, termed interaction parameter (I). A strong adsorption would be indicated by I approaching a value of 1·0. Thus, it would be convenient to compare the relative adsorption behavior of several surfactants at a given surface and also of a given surfactant at several surfaces. The interaction parameter approach is used to evaluate surfactant adsorption results obtained at PS and PVAc latex surfaces.

Ionic surfactant adsorption data for the PS and PVAc systems are summarized in Table III. The following points are worth noting. As mentioned earlier, the presennt values of A_0 and A_m for the three ionic surfactants at PS latex surface agree quite well with previous findings. In general, the adsorption of ionic surfactants at concentrations close to their respective CMCs is weaker at a PVAc latex surface compared to that at a PS latex surface. This is shown by larger A_m and smaller I values observed in the PVAc system compared to the PS system. It is seen that adsorption of ionic surfactants studied at the PS surface increases with the hydrophobicity of the surfactant. For instance, I values increase with increasing hydrophobicity of the surfactant as judged by their decreasing CMC values (see Table I). In the case of cationic

TABLE III
Interaction parameter of surfactants

	Area/molecule (nm²)				
		Area per molecule at CMC, A_m		*Interaction parameter I, A_0/A_m*	
Surfactant ionic	*Limiting area, A_0*	*Poly-styrene*	*Poly(vinyl acetate)*	*Poly-styrene*	*Poly(vinyl acetate)*
Sodium lauryl sulfate	0·26–0·30 (0·29)[a]	0·45 (0·45–0·55)[a]	1·1	0·63	0·26
Sodium oleate	0·22–0·25 (0·21)	0·30 (0·28)	2·74	0·71	0·08
Hexa decyl trimethyl ammonium bromide	0·35–0·37 (0·37)	0·47 (0·47)	1·33	0·77	0·28

[a] Literature values are given in parentheses

CTABr, the initial adsorption is due to charge neutralization of the anionic groups ($SO_4^=$ and COO^-) at the latex surface, resulting in initial flocculation and subsequent stabilization of latex particles due to charge reversal. In the PVAc system, the already weak adsorption observed with NaLS is further diminished as the hydrophobicity of the surfactant is increased in Na oleate. However, the I value for the more hydrophobic CTABr is found to be much higher (0·28) than the value observed for Na oleate (0·08). This apparent anomaly is due to the fact that most of the adsorption observed with CTABr on PVAc is due to the initial charge neutralization, leading to flocculation of the latex and subsequent charge reversal and dispersion. This observation is consistent with those of Connor and Ottewil.[10]

Thus, it appears that the adsorption of anionic surfactants at the more hydrophobic PS latex surface increases with increase in hydrophobicity of the surfactant while at the more hydrophilic PVAc latex surface, the adsorption decreases with increase in hydrophobicity of the anionic surfactant. If other competing reactions such as charge neutralization, complexation and solubilization are significant in the surfactant adsorption process the above observation on the effect of hydrophobicity of surfactant on adsorption may not be completely valid.

Similar results for the Igepal CO-type nonionic surfactants are shown in Table IV. The experimental A_0 and A_m values for the various Igepal CO surfactants studied could not be compared to literature values, as they are not

TABLE IV
Interaction parameter of surfactants

	Area per molecule (nm²)				
		Area per molecule at CMC, A_m		*Interaction parameter I, A_0/A_m*	
Surfactant nonionic	*Limiting area, A_0*	*Poly-styrene*	*Poly(vinyl acetate)*	*Poly-styrene*	*Poly(vinyl acetate)*
Igepal CO-710 HLB = 13·3	0·60	0·9	5·5	0·67	0·11
Igepal CO-850 HLB = 16	0·82	1·5	3·5	0·56	0·23
Igepal CO-880 HLB = 17	1·01	—	4·6	—	0·22
Igepal CO-970 HLB = 18·2	1·25	2·5	18·4	0·5	0·07

readily available. The following trends are noted. Again, it is seen that the adsorption of nonionics on the more hydrophilic PVAc surfaces is weaker than at the more hydrophobic PS latex surface, as shown by larger A_m and smaller I values in the PVAc system. With PS latex, the adsorption decreases with decrease in hydrophobicity of the surfactant (increasing HLB), in agreement with results found for anionic surfactants. With PVAc latex, the interaction parameter, I, increases initially with increase in hydrophobicity of the surfactant (HLB) and then decreases. Thus, it appears there is an optimum HLB range for maximum adsorption of nonionic surfactants at a PVAc latex surface. The optimum HLB ranges of 16–17 and 13–14 found from the present adsorption studies for the PVAc and PS latexes respectively, agree quite well with the optimum HLB values reported for PVAc and PS latexes by Greth and Wilson[12] from emulsion stability studies using mixed surfactant systems.

The surfactant adsorption results presented above can at least be qualitatively explained by considering the relative hydrophobic/hydrophilic nature of the surfactant and polarity of the latex surface. In the next section, adsorption data are presented to study the effect of latex polarity on surfactant adsorption in a more quantitative manner.

(b) Polarity of vinyl acrylic latex and adsorption of nonionic surfactants

Adsorption results discussed above clearly show that in general surfactants do not adsorb strongly at the hydrophilic (polar) vinyl acetate latex surface. The higher hydrophilicity of PVAc latex is due to the significant water solubility of vinyl acetate monomer. It is well known[13,14] that water solubility of an organic phase can affect interfacial properties such as interfacial tension at an organic liquid/water interface, solubilization, adsorption, etc. Paxton[5] has shown experimentally that the equilibrium area (A_m) occupied by a surfactant molecule at a latex/water interface changes with the nature of the latex composition. More recent work[15,16] from this laboratory has successfully correlated the area occupied by a surfactant molecule at an organic/water interface to the polarity of the organic phase. It was shown that water solubility of organic phase can be quantitatively related to polarity (ratio of polar contribution to the total surface free energy of the organic phase) and that the logarithm of area occupied by a surfactant molecule at an interface increases with the polarity of the organic phase. Figure 9, taken from reference 15, shows the effect of polymer polarity on surfactant adsorption, for NaLS at various latex/water interfaces.

In this section, the effect of polymer polarity on surfactant adsorption at model vinyl acrylic latex surfaces is studied. Vinyl acrylic latex of varying

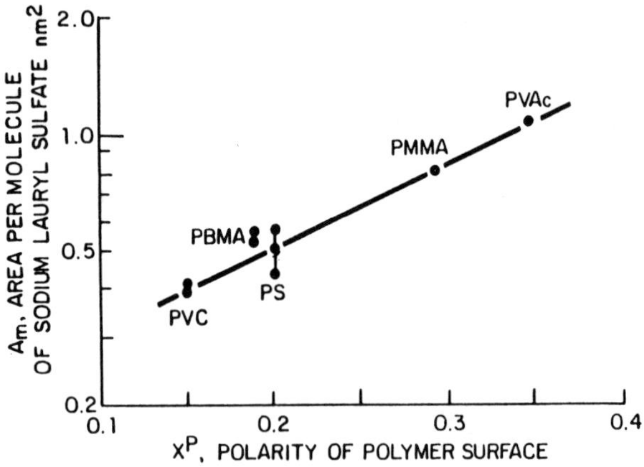

Fig. 9. Plot of $\log A_m$ versus polarity (X^P) for NaLS at various latex/water interfaces.[15]

polarity was prepared by changing the butyl acrylate (BA) content of the VAc/BA latexes. The latexes were 'cleaned' by dialysis and ion exchange. Polarity of the copolymer latex surface was calculated from contact angle measurements.

Contact angle measurements, dispersion and polar contribution to latex film surface tension and polarity of polymer, calculated according to the method of Kaelble,[7] of the three latex films are shown in Table V. It is seen that the polarity of the latex film decreases with increase in butyl acrylate content of the vinyl acrylic copolymer. The polarity of the 70/30 (VAc/BA) latex is very similar to that of poly(butyl acrylate), estimated to be about 0·21.[15]

TABLE V
Polarity of latex surface by contact angle measurements

Latex	*Contact angle, θ* H_2O	CH_2I_2	*Dispersion contributions* γ_d *(dynes/cm)*	*Polar contributions* γ_p *(dynes/cm)*	*Polarity* $X^P = \gamma_p/\gamma_p + \gamma_d$
VAc/BA 70/30	72	41	30·5	9·3	0·23
VAc/BA 85/15	67	38	30·0	12·2	0·29
VAc/BA 100/0	44	29	25·4	29·8	0·54

VAc = Vinyl acetate. BA = Butyl acrylate. θ = Average of eight measurements.

Polarity of 0·54 for the PVAc homopolymer is considerably higher than the 0·33 value determined for the homopolymer from interfacial tension measurements on polymer melts.[17] The higher polarity value found in the PVAc film strongly suggests that there was a significant amount of hydrolysis of PVAc during polymerization and/or cleaning of the latex resulting in a more hydrophilic PVAc surface.

Table VI shows the adsorption data of Igepal CO-630 surfactant at the three latex/water interfaces. The isotherms were of the normal type and no thickening of the latex in the presence of surfactant was observed (see Section (c)). Area per molecule was calculated according to eqn. (1).

TABLE VI
Adsorption of Igepal CO-630 on vinyl acrylic latex of different polarity

Latex	X^P *Polarity*	*D, Particle size (nm)*	*Adsorption (moles/cm² of polymer)*	*A, Area per molecule (nm²)*	*I, Interaction parameter*
VAc	0·54	0·057	$8{\cdot}89 \times 10^{-11}$	1·87	0·29
VAc/BA 85/15	0·29	0·062	$14{\cdot}7 \times 10^{-11}$	1·13	0·49
VAc/BA 70/30	0·23	0·073	$17{\cdot}3 \times 10^{-11}$	0·96	0·57

I = Interaction parameter = A_0/A
$A_0 = 0{\cdot}55$ nm² for Igepal CO-630

It is seen that the adsorption of Igepal CO-630 decreases with increase in polarity of the vinyl acrylic latex. Also the interaction parameter, as expected, decreases with increase in polarity of the latex surface. It shows that at saturation adsorption, the extent of interaction of Igepal CO-630 with the PVAc homopolymer and VAc/BA copolymer latexes under investigation is about 30% and 50%, respectively, of the theoretical limit corresponding to a close packed monolayer adsorption.

Figure 10 shows a plot of $\log A$ versus polarity (X^P) of latex surface. It is readily seen that the plot is quite linear and fits the following equation, derived in reference 15,

$$\log A = \text{const} + k(X^P) \tag{2}$$

Attempts to correlate the adsorption of other anionic surfactants such as Alipal EP-110 and NaLS on the three latex surfaces in a similar manner failed because of the more complex and specific interactions observed in these

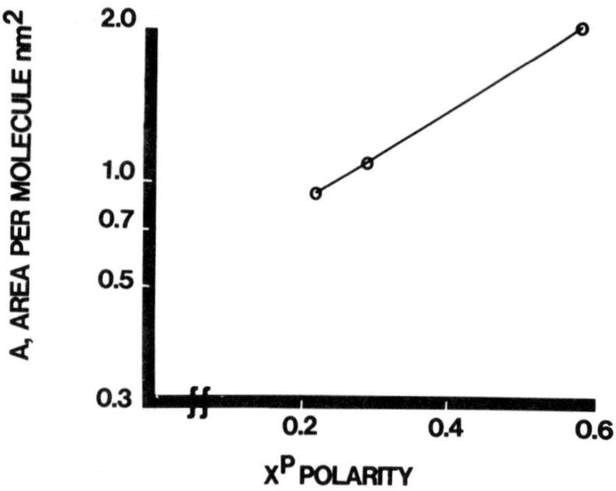

Fig. 10. Plot of log A versus polymer polarity at vinyl acrylic latex/water interface.

systems (see Section (c)). Equation (2) can adequately describe the adsorption data of surfactants at polymer/water interfaces, provided that the free energy of the interface is related to the free energy of adsorption and there are no specific interactions between surfactant and interface.[16]

The polarity and adsorption data discussed above reveal some interesting aspects of the surface chemistry of vinyl acrylic latex surfaces. It is quite likely that the polarity of the latex films, especially of the two copolymers, determined by contact angle measurements may not correspond exactly with their respective latex surfaces in the dispersed state, due to reorientation of polymer chains during film formation. But the surfactant adsorption data show clearly that the three latex surfaces in their dispersed state do exhibit varying polarity paralleling the trend found from contact angle measurements. The result also shows that the surface of the copolymer latex surface is a mixture of vinyl acetate and acrylate units. This result is somewhat unexpected in a vinyl acrylic latex, prepared by a batch process since one would expect the more reactive butyl acrylate to predominate in the core structure leaving the less reactive vinyl acetate near the latex surface. Based on the electron micrographs of batch and semi-continuous polymerized 63/37 VAc/BA latex particles, Misra *et al.*[18] concluded that the batch sample has a heterogeneous structure comprising a relatively large butyl-acrylate-rich core surrounded by a vinyl-acetate-rich shell. In comparison, the semi-continuous polymerized particles were found to have a homogeneous structure comprised of a much smaller butyl-acrylate-rich core surrounded by a shell of vinyl acetate–butyl acrylate copolymer. It would be interesting to study the two VAc/BA latex samples of Misra *et al.* by the above methods to ascertain the differences in the surface composition of the two latexes.

Our investigations on the effect of polarity and surfactant adsorption generally shows that the extent of surfactant adsorption and energy of adsorption at an organic liquid/water interface decreases with increase in polarity (increased water solubility) of the organic phase. Our findings are in excellent agreement with the results of Yeliseyvea and Zuikov[14] who have carried out extensive investigations on the surfactant adsorption behavior at alkyl acrylate/water interfaces. They have found, as shown in Table VII, that with increasing water solubility (polarity) of the alkyl acrylate, saturation adsorption of NaLS at the monomer/water and polymer/water interface decreases and so does the energy of adsorption. They have also related the energy of adsorption to an empirical rate equation of the form.

$$\text{Rate} = \text{Const.}\,(C_{\text{emuls}})^{X}\,(C_{\text{init}})^{0\cdot5} \qquad (3)$$

X has been found to increase with energy of adsorption as shown in Table VIII.

Zuikov *et al.*[19] have reported that the energy of adsorption of an emulsifier at a monomer/water interface can be used to select the optimum emulsifier(s) for a given latex composition. This appears to be an interesting approach for surfactant selection and merits further investigation.

To summarize this section, we have shown that:

(a) Adsorption of nonionic surfactants at a vinyl acrylic copolymer latex/water interface can be correlated to the polarity of the vinyl acrylic latex, a result in agreement with previous work on the adsorption of NaLS at homopolymer latex/water and organic liquid/water interfaces.

(b) Contact angle measurement and surfactant adsorption can be used to characterize the polarity of copolymer latex surfaces.

TABLE VII

Adsorption of sodium lauryl sulfate at alkyl acrylate/water interface

Alkyl acrylate	*Water solubility of monomer (wt%)*	*Monomer/ water interfacial tension (dynes/cm)*	*Energy of adsorption (kcal/mole)*	*Area per molecule (nm^2)*
Methyl	5·69	13·0	5·1	1·920
Ethyl	1·84	21·5	5·6	1·238
Butyl	0·34	28·5	6·5	1·057

Yeliseyvea and Zuikov[14]

TABLE VIII
Effect of energy of adsorption of emulsifier at monomer/water interface on polymerization rate

Emulsifier	*Energy of adsorption at ethyl acrylate/water (kcal/mole)*	*X*
Sodium lauryl sulfate	5·6	0·21
Alkylpolyglycol sulfate (10 moles of EO)	5·9	0·25
Sulfated Triton X-100	6·5	—
C-30 Polyglycol ether of cetyl alcohol (30 moles EO)	6·4	0·38

$R = K\,[C_{\text{emuls}}]^{X}\,[C_{\text{init}}]^{0.5}$

Yeliseyvea and Zuikov[14]

(c) Specific interactions of anionic surfactants in vinyl acrylic latex systems

Recent investigations[20,21] have shown that the behavior and interactions of anionic surfactants in poly(vinyl acetate) latex are quite different and complex compared to that in a polystyrene latex. Earlier work[22–24] has also shown that anionic surfactants adsorb on poly(vinyl acetate), then slowly penetrate into the particle, leading to the formation of a polyelectrolyte type solubilized polymer–surfactant complex. Such a solubilization process is generally accompanied by an increase in viscosity.

In this section, the effects of type and structure of anionic surfactants on the solubilization phenomena in vinyl acetate and vinyl acetate–butyl acrylate copolymer latex are considered. The anionic surfactants investigated are NaLS, Aerosol A-102, Alipal EP-110 and Alipal EP-120.

Adsorption isotherms of NaLS and Aerosol A-102 surfactants on a 'clean' 85/15 VAc/BA copolymer latex are shown in Fig. 11. It is seen that the shapes of the two isotherms are quite different. A-102 seems to exhibit a Langmuir type saturation adsorption behavior typical of surfactant adsorption at a polystyrene/water interface. NaLS seems to have a more complex adsorption isotherm, perhaps a multiple step process as discussed by Giles *et al.*[25] The initial region of the isotherm up to B appears to be normal and beyond that the isotherm becomes linear. Giles *et al.* have explained such a linear isotherm (classified as Type C) by penetration of substrate by solute, leading to new adsorption sites. Compared to NaLS, the size of Aerosol A-102 molecule is much larger and bulkier and thus perhaps hindered in being able to penetrate into the latex particle.

Figure 12 shows the adsorption isotherms of two sulfated ethyoxylate

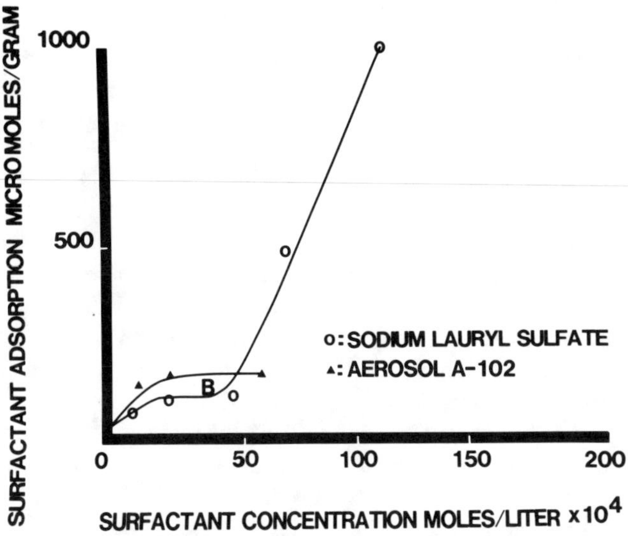

Fig. 11. Adsorption isotherm of NaLS and Aerosol A-102 surfactants at 85/15 VAc/BA copolymer latex/water interface.

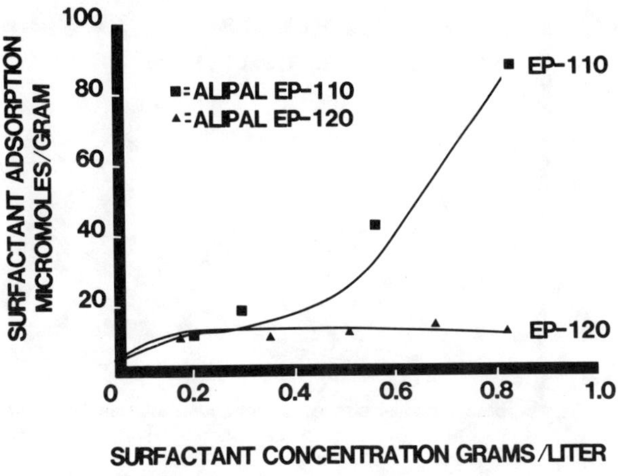

Fig. 12. Adsorption isotherm of Alipal EP-110 and Alipal EP-120 at 85/15 VAc/BA copolymer latex/water interface.

type anionics–Aliphal EP-110 and Alipal EP-120–on the 85/15 VAc/BA latex surface. Again, it is seen that the lower molecular weight EP-110 shows a C type isotherm similar to NaLS while the higher molecular weight EP-120 exhibits a normal saturation type isotherm.

It is well known[22–24] that sodium lauryl sulfate interacts with some polymers such as poly(vinyl acetate) causing solubilization of the insoluble polymer leading to an increase in viscosity. In Fig. 13, viscosity of the homopolymer and 70/30 VAc/BA at various NaLS/polymer ratios is shown. It is seen that the viscosity of the 2% latex dispersion increases with increase in NaLS/polymer ratio. Similar viscosity data for the 85/15 VAc/BA were intermediate between the homopolymer and 70/30 VAc/BA latexes. Surfactants that showed a normal saturation type adsorption behavior did not show any significant viscosity increase of the latex.

Edelhauser[24] has explained his results on the interaction of anionic surfactants (sodium lauryl sulfate and sodium dodecylbenzene sulfonate) with a PVAc latex as a two-step mechanism involving surface adsorption followed by surfactant penetration into the interior of PVAc particles taking water with it and causing swelling and gradual dissolution of the disintegrated polymer chains. Based on this model, it is convenient to classify the adsorption behavior of anionic surfactants at vinyl acrylic latex surface into penetrating

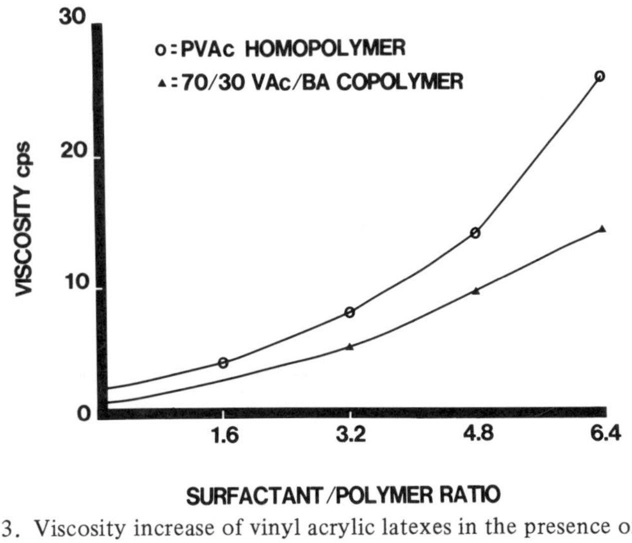

Fig. 13. Viscosity increase of vinyl acrylic latexes in the presence of NaLS.

and non-penetrating types as shown in Table IX. The smaller surfactant molecules seem to be capable of penetrating into the latex, causing disintegration and solubilization of the otherwise insoluble polymer. The larger and bulkier surfactants studied seem to adsorb only at the latex surface, and are unable to penetrate into the particle. It appears that there exists a critical size and perhaps shape of anionic surfactant molecule, for the surfactant to be able to penetrate into the particle. It is also likely that the nature of hydrophobe adsorbing at the latex/water interface and the charge density on the polymer–surfactant complex may influence surfactant penetration and latex solubilization. More extensive work with other anionic surfactants of varying chain length and shape would be needed to better define molecular parameters that influence the penetration of surfactants into a vinyl acrylic latex.

Latex thickening in the presence of penetrating type anionic surfactants such as NaLS appears to depend on polymer composition as seen in Table X.

TABLE IX
Anionic surfactants studied in vinyl acrylic latexes

Penetrating type

(A) Sodium lauryl sulfate
$C_{12}H_{25}SO_4Na$ M wt = 288·4

(B) Sodium dodecyl benzene sulfonate

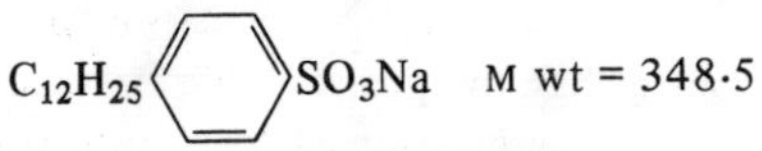

(C) Alipal EP-110 M wt = 708

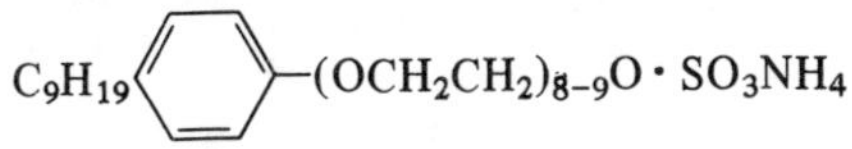

Non-penetrating type

(A) Aerosol A-102 M wt = 1800
disodium ethoxylated alcohol (C_{10}–C_{12}) half ester of sulfosuccinic acid

$CH_2 \cdot COO(H_2C \cdot CH_2O)_{30}C_{11} \cdot H_{23}$
|
$CHCOONa$
|
SO_3Na

(B) Alipal EP-120 M wt = 1640

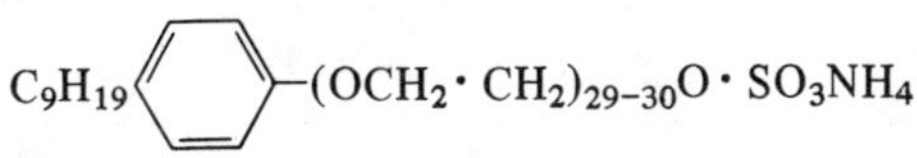

TABLE X
Effect of latex composition on thickening by sodium lauryl sulfate

Latex composition	*Latex thickening*
Poly(vinyl acetate)	19·3
Poly(vinyl acetate)–butyl acrylate (70/30)	10·0
Polystyrene	1·1
Poly(methyl methacrylate–ethyl acrylate)	1·1

$$\text{Thickening } \frac{\eta}{\eta_0} = \frac{\text{Brookfield viscosity of 2\% latex in the presence of NaLS}}{\text{Brookfield viscosity of 2\% control latex}}$$

The extent of latex thickening in the presence of excess NaLS decreases with decrease in the VAc content of a vinyl acetate–butyl acrylate copolymer. Polystyrene and poly(methacrylate–acrylate) copolymer latexes do not show any thickening.

The presence of nonionic surfactants such as Igepal CO-630 seems to prevent the thickening of PVAc latex by NaLS, as shown in Table XI. This can be interpreted as showing that the presence of nonionic surfactant at the PVAc latex surface prevents the penetration of NaLS into the particle.

TABLE XI
Effect of nonionic surfactant on thickening of poly(vinyl acetate)–sodium lauryl sulfate (NaLS)

Latex	*Latex thickening*†
Poly(vinyl acetate) + NaLS	19·3
Poly(vinyl acetate) + Igepal CO-630	1·0
Post addition of Igepal CO-630 to poly(vinyl acetate) + NaLS	17·4
Post addition of NaLS to poly(vinyl acetate) + Igepal CO-630	1·3
Poly(vinyl acetate) + mixed Igepal CO-630 and NaLS	1·4

† See Table X for definition of latex thickening

To summarize this section, the following points are noted:

(a) Lower molecular weight (300–800) anionic surfactants such as NaLS and Alipal EP-110 adsorb at a vinyl acrylic latex in a Complex C-type adsorption isotherm, leading to penetration and solubilization of latex.
(b) Relatively higher molecular weight (1600) anionic surfactants such as Alipal EP-120 and Aerosol A-102 adsorb in a normal manner, suggesting that there exists a critical size and perhaps shape for the surfactant to be able to penetrate into the latex.
(c) Latex thickening, a result of surfactant penetration and solubilization of the polymer, decreases with decrease in vinyl acetate content of the vinyl acetate–butyl acrylate copolymer.
(d) Presence of nonionic surfactants at the latex/water interface seem to interfere with the penetration of low molecular weight anionic surfactants.

(d) Effect of surfactant on film formation of vinyl acrylic latexes

Previous studies[26–28] have shown that surfactants influence the film formation process in latexes. For instance, Bradford and Vanderhoff[26] have shown that during the film formation process, surfactants that are not compatible with the base latex polymer exude out into a separate phase and form domains of two-phase regions. Their work also showed that with hydrophobic latexes such as styrene–butadiene, the more hydrophobic lower HLB nonionic surfactants were more compatible and formed more uniform films than the latexes containing higher HLB surfactants. SBR latexes prepared with the more hydrophilic higher HLB surfactants formed a separate 'exudate' phase due to incompatibility. Further, they showed that the presence of adsorbed stabilizer layers at the latex interface could influence the further gradual coalescence phenomena observed in latex film formation process.

In this section, we report some work on the effect of surfactants on the film formation process in vinyl acrylic latexes. Table XII shows the effect of NaLS surfactant on the glass transition temperature (*Tg*) of PVAc homopolymer, VAc/EA and VAc/BA copolymer latexes. It is readily seen that NaLS acts as a plasticizer and lowers the *Tg* of the latexes.

Table XIII shows the effect of HLB of nonionic surfactants on *Tg* and MFFT (minimum film formation temperature) of a VAc/BA latex. It is seen that the higher HLB (more hydrophilic) surfactant has a stronger effect in lowering the *Tg* and MFFT compared to the lower HLB one. This shows that the higher HLB nonionic surfactant is more compatible with the fairly hydrophilic vinyl acrylic latex. As mentioned above, the more hydrophobic lower HLB surfactants were found to be more compatible with a hydro-

TABLE XII
Effect of sodium lauryl sulfate on glass transition temperature (*Tg*)

Latex	*Glass transition temperature* °C *Before dialysis*	*After dialysis*
Poly(vinyl acetate)	28	42
Poly(vinyl acetate)–ethyl acrylate	18	27
Poly(vinyl acetate)–butyl acrylate	13	17

TABLE XIII
Effect of nonionic surfactant on glass transition temperature (*Tg*) and minimum film formation temperature (MFFT) of vinyl acetate–butyl acrylate latex

Surfactant type	*Surfactant level*[a] (%)	*Tg* (°C)	*MFFT* (°C)
Igepal CO-850 (HLB = 16)	3	26	7
CO-850	5	24	6
Igepal CO-977 (HLB = 18)	3	24	4
CO-977	5	16	3

[a] Based on monomer

phobic SBR latex. These results are in good agreement with our findings on the effect of HLB on surfactant adsorption at PVAc and PS latex surfaces.

Transmission electron microscopy of surface replicas of latex films shadowed with Pt/Pd have also been studied to investigate the effects of surfactants on the film formation process in vinyl acrylic latexes. Results of this study are discussed in detail elsewhere.[29] Here, some typical electron micrographs of latex films are shown.

Figure 14 shows the effect of NaLS on the surface morphology of a 15 day aged sample of PVAc latex film. It is seen that in the presence of NaLS, the surface has a different structure and appears to be formed of a separate phase. This is much more clearly evident in the case of VAc/BA copolymer latex film (Fig. 15(A)), where the top surface seems to have a different morphology. In the absence of NaLS, the copolymer seems to form a smooth uniform film (Fig. 15(B)), while in the homopolymer sample the latexes are not completely coalesced (Fig. 14(B)). Figure 16 shows the effect of HLB of

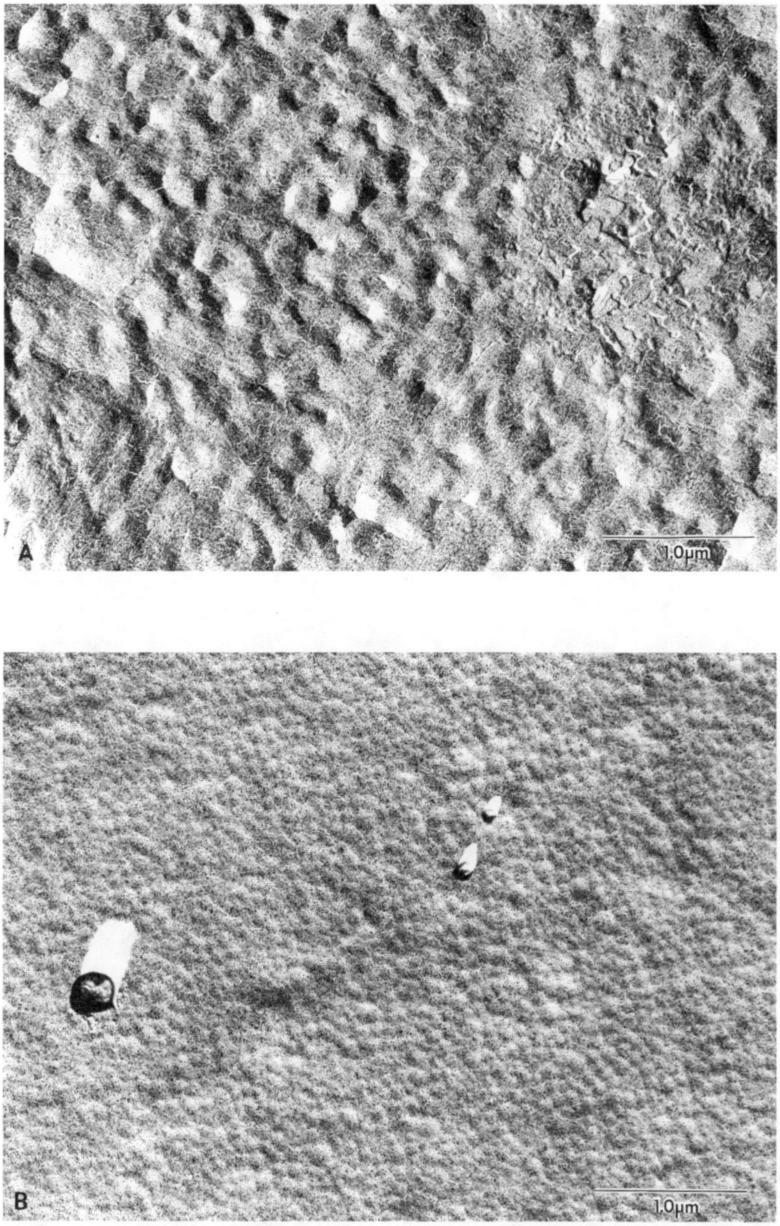

Fig. 14. Electron micrographs of surface replicas of PVAc latex films in the presence (A) and absence (B) of NaLS. Films aged 15 days. × = 32 250.

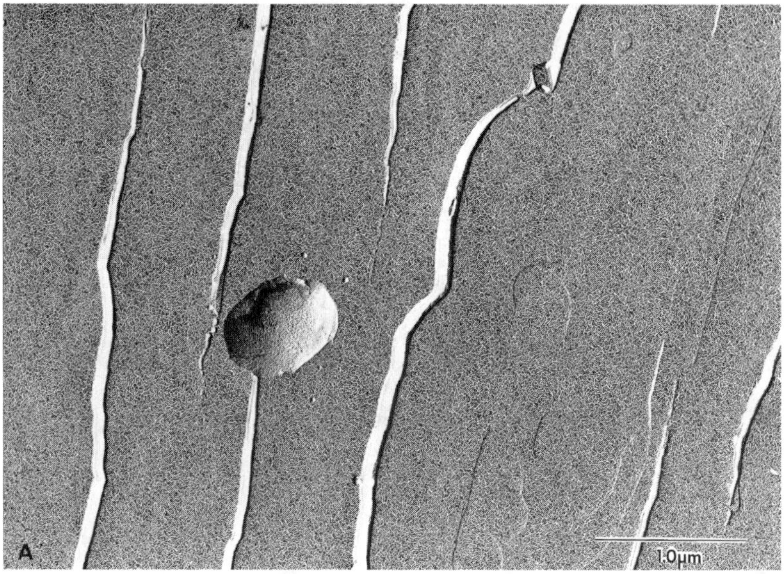

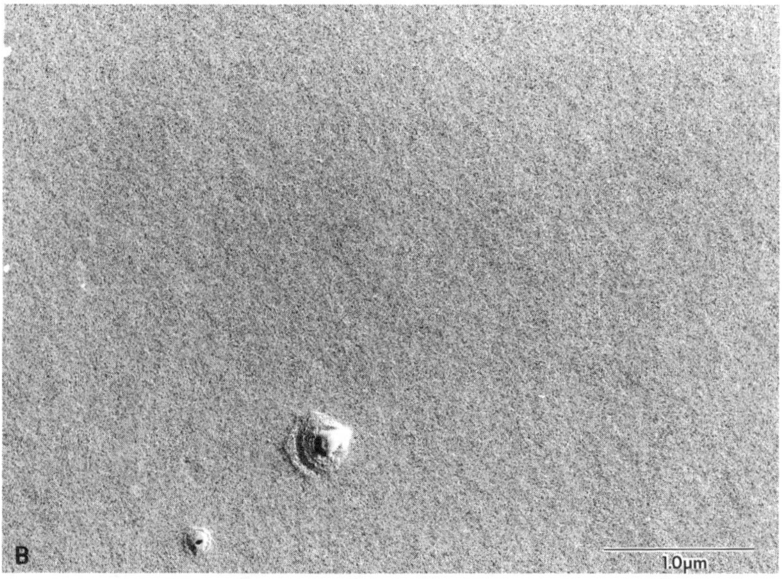

Fig. 15. Electron micrographs of surface replicas of VAc/BA copolymer latexes in the presence (A) and absence (B) of NaLS. Films aged 15 days. $\times = 32\,250$.

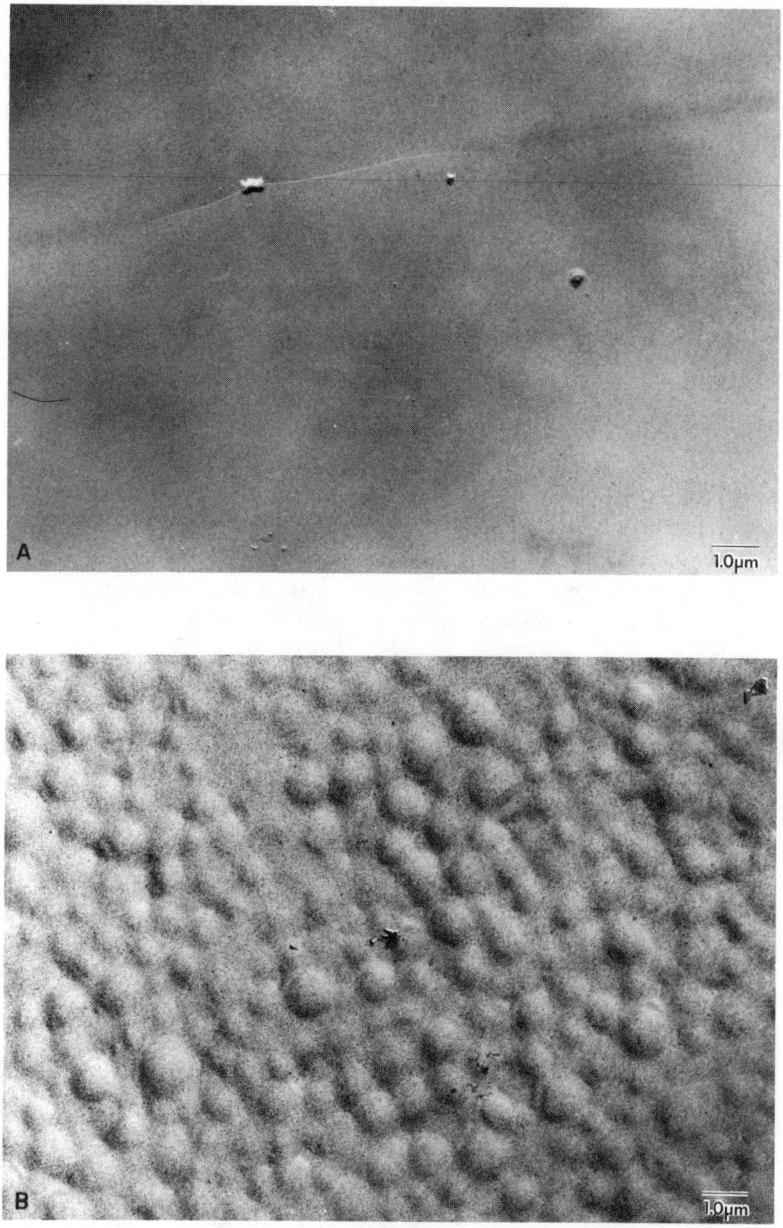

Fig. 16. Electron micrographs of surface replicas of VAc/BA latex in the presence of Igepal CO-977 (A) and Igepal CO-850 (B) nonionic surfactants.

nonionic surfactants on film coalescence of a VAc/BA latex. Latex containing the higher HLB Igepal CO-977 surfactant seems to form a smoother film (Fig. 16(A)) compared to the one containing the lower HLB Igepal CO-850 surfactant (Fig. 16(B)). This observed influence of nonionic surfactants on film formation is in agreement with the data on the effect of HLB of nonionic surfactants on *Tg* and MFFT (Table XIII) and adsorption at the polar vinyl acrylic latex surface as discussed in Section (a).

The effect of NaLS on latex film morphology (Figs. 14 and 15) can be explained as follows. Vinyl acrylic latex forms a solubilized complex with NaLS (as discussed in the previous section) and the fairly hydrophilic complex being incompatible with the base polymer exudes out into a separate phase. It is also likely that the solubilized complex may be pushed to the surface by the convectional gradients set up during the film formation process due to hydrodynamic and surface tension forces. A recent study[30] has shown that presence of smaller particles and low surface tension in a latex system could strongly contribute to pushing of the smaller particles to the surface. Such a mechanism may also be operative here. In the absence of NaLS, the homopolymer forms a poorly coalesced film showing that the further gradual coalescence phenomena observed in SBR and all acrylic latexes is not operative in PVAc latexes. This could be due to the presence of adsorbed and grafted polyvinyl alcohol layers (see next section) around the PVAc particles acting as steric barrier to the interpolymer chain penetration, as suggested by Bradford and Vanderhoff.[26]

Surfactant effects on the film formation of vinyl acrylic latexes can be summarized as follows.

(a) Surfactants–nonionic and anionic–generally lower the *Tg* and MFFT of vinyl acrylic latexes. Surfactants that are more compatible with the polymer have a stronger effect.
(b) Latex film morphology is influenced by the nature of the surfactant and possible interaction complexes formed between the polymer and surfactant. An example of this effect is that of NaLS and NaLS–vinyl acrylic latex complex on film formation and morphology.
(c) In PVAc homopolymer latex, the further gradual coalescence phenomenon is not significant. This could be due to the presence of hydrolysed polyvinyl alcohol layers around the latex particles acting as steric barriers to the interpretation of polymer chains.

(e) Grafting of nonionic surfactants and stabilizers in vinyl acetate polymerization

In this section, we review another of the distinguishing features of the role of nonionic surfactants and stabilizers in the emulsion polymerization of vinyl acetate, namely grafting through chain transfer reactions. Okamura and Motoyama,[1] in comparing the polymerization kinetics of vinyl acetate and styrene in the presence of a nonionic surfactant, polyethylene oxide dodecyl ether, note that the chain transfer constant for vinyl acetate is 0·075 and only 0·002 for styrene. They also observed that the rate of polymerization for vinyl acetate may even decrease as the concentration of nonionic surfactant is increased. Lunardon *et al.*[31] studied the emulsion polymerization of vinyl acetate in the presence of nonylphenol-polyoxyethylene adducts of various chain lengths and found the rate of polymerization to be proportional to the 0·25 power of surfactant concentration and 0·8 power of initiator concentration. These and other studies clearly show that polymerization of vinyl acetate does not follow the classical Smith–Ewart polymerization kinetics.

In a study of the polymerization of vinyl acetate in presence of varying amounts of polyvinyl alcohol as nonionic stabilizer, Hartley[32] found evidence for the presence of poly(vinyl acetate)–polyvinyl alcohol graft copolymer. Two different species of graft copolymer, one water-soluble and of high polyvinyl alcohol content, and the other water-insoluble and of low polyvinyl alcohol content were isolated by Hartley through chromatographic fractionation. The presence of an infrared absorption band at 8·8 μm, a characteristic of a tertiary alcohol group, in both species of graft copolymer led Hartley to suggest that grafted poly(vinyl acetate) chains are attached to the chain carbon atoms of the polyvinyl alcohol molecules to which the hydroxyl groups are also attached. Stepin and Tolmacher[33] confirmed the IR evidence of grafting and claimed that treatment with oxidizing agents destroys the grafted polyvinyl alcohol and thus destabilizes the aqueous poly(vinyl acetate) dispersion. Traeen,[34] however, using techniques similar to that of Hartley, did not find any evidence of the grafting of polyvinyl alcohol to vinyl acetate.

Gavat *et al.*[35] have also investigated the grafting of nonionic surfactants and stabilizers in the vinyl acetate emulsion polymerization process. They observed a decrease in the polymerization rate of vinyl acetate as polyvinyl alcohol concentration was increased and attributed this to the grafting process which results in macroradicals whose activity is lower than that of poly(vinyl acetate) radicals. Gavat *et al.* state that an important parameter of the grafting reactions is the polyvinyl alcohol/vinyl acetate ratio. In a comparative study, with ethylene oxide–propylene oxide (EO–PO) block copolymer as nonionic emulsifiers (with 80 wt% EO units), Gavat *et al.*[35] found that grafting of vinyl acetate occurred on the EO–PO block copolymer chains also. In fact, their data suggest that more grafting takes place with EO–PO block copolymers which may prove to be stronger transfer agents than polyvinyl alcohol. The presence of the grafted copolymers was determined by IR absorption bands at 805–807 cm^{-1}, which is taken as characteristic of all of the grafted copolymers.

A few studies have examined the colloidal stability of PVAc latexes prepared in the presence of surfactants and protective colloids that have active sites for grafting reactions. For instance, Stepin and Tolmacher[33] found that 'cleaned' PVAc latexes (repeated washing until a negative reaction for polyvinyl alcohol in the wash water was obtained) prepared in the presence of polyvinyl alcohol did not coagulate under extreme temperatures (80°C for 10 hours and −40°C for 5 hours) and high electrolyte concentration. They attributed the observed remarkable stability of PVAc latexes to the grafting of polyvinyl alcohol to vinyl acetate at the surface of the polymer particles. They also reported that the latex, after treatment with oxidizing agents to destroy the grafted polyvinyl alcohol chain, lost its stability.

Preliminary work done in our laboratory,[36] using the serum replacement technique developed by the University of Lehigh group,[9] shows that it is not possible to remove a significant amount of Igepal CO type (nonylphenol–ethylene oxide adduct) nonionic surfactant from poly(vinyl acetate) latex surface even after extensive washing. This suggests that at least some of the nonionic surfactant is attached to the latex particle surface by a mechanism[9] other than physical adsorption force.

As the brief discussion here shows, most of the available literature data appear to suggest the formation of grafted copolymer when vinyl acetate is polymerized in presence of nonionic surfactant and polymeric stabilizer. However, further careful and systematic work is needed to better understand the chain transfer tendency of such nonionic stabilizers in vinyl acrylic latexes.

ACKNOWLEDGEMENT

It is a pleasure to thank Celanese Plastics and Specialties Company for permission to publish this work.

REFERENCES

1. S. Okamura and T. Motoyama, *J. Polym. Sci.*, **58**, 221 (1962).
2. B. R. Vijayendran, *J. Coll. and Int. Sci.*, **60**, 418 (1977).
3. H. J. Van den Hull and J. W. Vanderhoff, *Br. Polym. J.*, **2**, 121 (1970).
4. S. H. Maron, M. E. Elder and I. N. Ulevitch, *J. Coll. and Int. Sci.*, **3**, 89 (1954).

5. T. R. Paxton, *J. Coll. and Int. Sci.,* **31**, 19 (1969).
6. J. R. Epton, *Trans. Faraday Soc.,* **64**, 226 (1948).
7. D. H. Kaelble, *Physical Chemistry of Adhesion,* Wiley-Interscience, New York, 1971, Chapt. 4.
8. H. J. Van den Hull and J. W. Vanderhoff, in *Polymer Colloids,* R. M. Fitch (ed.), Plenum Press, New York, 1971.
9. S. M. Ahmed, M. S. El-Aasser, F. J. Micale, G. W. Poehlein and J. W. Vanderhoff, *J. Coll. and Int. Sci.,* **73**, 388 (1980).
10. P. Connor and R. H. Ottewil, *J. Coll. and Int. Sci.,* **37**, 642 (1971).
11. C. L. Sieglaff and J. Mazur, *J. Coll. Sci.,* **17**, 66 (1962).
12. G. G. Greth and J. E. Wilson, *J. Appl. Polym. Sci.,* **14**, 135 (1961).
13. C. O. Timmons and W. A. Zisman, *J. Coll. and Int. Sci.,* **28**, 106 (1968).
14. V. I. Yeliseyvea and A. V. Zuikov, in *Emulsion Polymerization, ACS Symposium Series,* Vol. 24, I. Piirma and J. L. Gordon (eds.), 1976, p. 62.
15. B. R. Vijayendran, *J. Appl. Polym. Sci.,* **23**, 733 (1979).
16. B. R. Vijayendran, *J. Coll. and Int. Sci.,* **68**, 383 (1979).
17. S. Wu, *J. Macromol. Sci.–Rev. Macromol. Sci.,* **C-10**, 1 (1974).
18. S. C. Misra, C. Pichot, M. S. El-AAsser and J. W. Vanderhoff, *J. Polym. Sci., Polym. Lett.,* **17**, 567 (1979).
19. A. V. Zuikov, A. I. Vassilenko and V. I. Yeliseyva, *Vyosomol Soedin, Series B,* **18**, 707 (1976).
20. B. R. Vijayendran, *Colloid Div. Abstracts,* Spring ACS Meeting, Honolulu, 1979.
21. A. M. Ahmed, M. S. El-Aasser, G. W. Poehlein and J. W. Vanderhoff, Emulsion Polymers Institute Research Reports, Lehigh University, Pennsylvania, 1979.
22. T. Isemura and A. Imanishi, *J. Polym. Sci.,* **33**, 337 (1950).
23. H. Arai and S. Harin, *J. Coll. and Int. Sci.,* **30**, 312 (1969).
24. H. A. Edelhauser, *J. Polym. Sci., Part C,* **27**, 291 (1969).
25. C. H. Giles, D. Smith and A. Huitson, *J. Coll. and Int. Sci.,* **47**, 755 (1974).
26. E. B. Bradford and J. W. Vanderhoff, *J. Macromol. Sci.–Phys.,* **B6**, 671 (1972).
27. E. B. Bradford and J. W. Vanderhoff, *J. Macromol. Chem.,* **1**, 335 (1966).
28. S. S. Voyutski and Z. M. Ustionova, *J. Adhesion,* **9**, 39 (1977).
29. B. R. Vijayendran, T. Bone and L. S. Sawyer, to be published.
30. D. P. Durbin, J. W. Vanderhoff and M. S. El-Aasser, Emulsion Polymers Institute Research Reports, Lehigh University, Pennsylvania, 1980.
31. G. F. Lunardon, G. P. Talamini and U. Grosso, *Europ. Polym. J.,* **11**, 437 (1975).
32. F. D. Hartley, *J. Polym. Sci.,* **34**, 397 (1959).
33. S. N. Stepin and I. A. Tolmacher, *Kollodiyni Zh.,* **37**, 1012 (1975). (English translation.)
34. A. H. Traeen, *J. Appl. Polym. Sci.,* **7**, 581 (1963).
35. T. Gavat, V. Dinonie, D. Donescu, C. Hagiopol, M. Munteanu, K. Gosa and Th. Deleanu, *J. Polym. Sci., Polym. Symp.,* **64**, 125 (1978).
36. B. R. Vijayendran and C. Gajria, Unpublished results, 1979.

Index